Cloning

OTHER BOOKS OF RELATED INTEREST

OPPOSING VIEWPOINTS SERIES
America Beyond 2001
Animal Rights
Biomedical Ethics
Endangered Species
The Environment
Global Resources
21st Century Earth

CURRENT CONTROVERSIES SERIES
Conserving the Environment
Ethics
Hunger
The Rights of Animals

AT ISSUE SERIES
Animal Experimentation
Cloning
Environmental Justice
Food Safety

Cloning

Lisa Yount, *Book Editor*

David L. Bender, *Publisher*
Bruno Leone, *Executive Editor*
Bonnie Szumski, *Editorial Director*
David M. Haugen, *Managing Editor*
Brenda Stalcup, *Series Editor*

Contemporary Issues
Companion

Greenhaven Press, Inc., San Diego, CA

Library of Congress Cataloging-in-Publication Data

Cloning / Lisa Yount, book editor.
    p.   cm. — (Contemporary issues companion)
    Includes bibliographical references and index.
    ISBN 0-7377-0330-X (lib. : alk. paper). —
ISBN 0-7377-0329-6 (pbk. : alk. paper)
    1. Cloning. 2. Cloning—Social aspects. 3. Cloning—Moral and
ethical aspects. I. Yount, Lisa. II. Series

QH442.2 .C56 2000
571.8'9—dc21   99-049733
                                                        CIP

©2000 by Greenhaven Press, Inc.
P.O. Box 289009, San Diego, CA 92198-9009

Printed in the U.S.A.

CONTENTS

Foreword   7

Introduction   9

Chapter 1: The Science of Cloning

1. Dolly, the Amazing Sheep Clone   13
   Michael Specter and Gina Kolata

2. Could Dolly Be a Hoax?   22
   Michael Waldholz

3. The Real Importance of Dolly   26
   Diana Lutz

4. Cloning Monkeys: Are Humans Next?   30
   Virginia Morell

5. Stem Cells: A Promising Line of Cloning-Related Research   39
   Gregg Easterbrook

Chapter 2: Cloning Animals

1. Cloning and the Livestock Industry   48
   Takahashi Seiya

2. "Pharming" Cloned Animals   53
   Elizabeth Pennisi

3. Cloned Animals May Suffer   59
   Meg Gordon

4. A Need for Genetic Diversity   63
   Keay Davidson

5. Can Cloning Save Endangered Species?   66
   Jon Cohen

6. Cloning an Extinct Animal   71
   Richard Stone

Chapter 3: Cloning and the Media

1. Cloning Facts and Fictions   79
   Jon Turney

2. "Cloning" Movie Scripts   84
   Lisa Bannon and Frederick Rose

3. The Messages of *Jurassic Park*   88
   Raymond G. Bohlin

4. Ian Wilmut: A Shy Scientist Weathers a Storm of Publicity   95
   Robert Lee Hotz

5. Richard Seed: Media Attention Feeds the Claims of a
   Publicity Seeker   101
   Dirk Johnson

Chapter 4: Human Cloning and the Law

1. Few Legal Barriers Prevent Human Cloning   106
   Margaret A. Jacobs
2. A Ban on U.S. Funds for Human Cloning Research   108
   Marlene Cimons and Jonathan Peterson
3. The National Bioethics Advisory Commission's Evaluation
   of Human Cloning   112
   Harold T. Shapiro
4. Is a Ban on Human Cloning Constitutional?   118
   Mark D. Eibert
5. Regulating Human Cloning   125
   Gregory E. Pence

Chapter 5: Should Humans Ever Be Cloned?

1. Ethical Issues Concerning Human Cloning   129
   Wray Herbert, Jeffery L. Sheler, and Traci Watson
2. Judeo-Christian Objections to Human Cloning   135
   Stephen G. Post
3. Cloning Could Halt Human Evolution   141
   Michael Mautner
4. Why Not Clone Humans?   144
   Robert Wachbroit
5. The Medical Benefits of Human Cloning   153
   The Human Cloning Foundation
6. Answers to Religious Arguments Against Human Cloning   157
   Ronald A. Lindsay

Glossary   162

Organizations to Contact   164

Bibliography   168

Index   171

FOREWORD

In the news, on the streets, and in neighborhoods, individuals are confronted with a variety of social problems. Such problems may affect people directly: A young woman may struggle with depression, suspect a friend of having bulimia, or watch a loved one battle cancer. And even the issues that do not directly affect her private life—such as religious cults, domestic violence, or legalized gambling—still impact the larger society in which she lives. Discovering and analyzing the complexities of issues that encompass communal and societal realms as well as the world of personal experience is a valuable educational goal in the modern world.

Effectively addressing social problems requires familiarity with a constantly changing stream of data. Becoming well informed about today's controversies is an intricate process that often involves reading myriad primary and secondary sources, analyzing political debates, weighing various experts' opinions—even listening to firsthand accounts of those directly affected by the issue. For students and general observers, this can be a daunting task because of the sheer volume of information available in books, periodicals, on the evening news, and on the Internet. Researching the consequences of legalized gambling, for example, might entail sifting through congressional testimony on gambling's societal effects, examining private studies on Indian gaming, perusing numerous websites devoted to Internet betting, and reading essays written by lottery winners as well as interviews with recovering compulsive gamblers. Obtaining valuable information can be time-consuming—since it often requires researchers to pore over numerous documents and commentaries before discovering a source relevant to their particular investigation.

Greenhaven's Contemporary Issues Companion series seeks to assist this process of research by providing readers with useful and pertinent information about today's complex issues. Each volume in this anthology series focuses on a topic of current interest, presenting informative and thought-provoking selections written from a wide variety of viewpoints. The readings selected by the editors include such diverse sources as personal accounts and case studies, pertinent factual and statistical articles, and relevant commentaries and overviews. This diversity of sources and views, found in every Contemporary Issues Companion, offers readers a broad perspective in one convenient volume.

In addition, each title in the Contemporary Issues Companion series is designed especially for young adults. The selections included in every volume are chosen for their accessibility and are expertly edited in consideration of both the reading and comprehension levels of the audience. The structure of the anthologies also enhances accessibility. An introductory essay places each issue in context and provides helpful facts such as historical background or current statistics and legislation that pertain to the topic. The chapters that follow organize the material and focus on specific aspects of the book's topic. Every essay is introduced by a brief summary of its main points and biographical information about the author. These summaries aid in comprehension and can also serve to direct readers to material of immediate interest and need. Finally, a comprehensive index allows readers to efficiently scan and locate content.

The Contemporary Issues Companion series is an ideal launching point for research on a particular topic. Each anthology in the series is composed of readings taken from an extensive gamut of resources, including periodicals, newspapers, books, government documents, the publications of private and public organizations, and Internet websites. In these volumes, readers will find factual support suitable for use in reports, debates, speeches, and research papers. The anthologies also facilitate further research, featuring a book and periodical bibliography and a list of organizations to contact for additional information.

A perfect resource for both students and the general reader, Greenhaven's Contemporary Issues Companion series is sure to be a valued source of current, readable information on social problems that interest young adults. It is the editors' hope that readers will find the Contemporary Issues Companion series useful as a starting point to formulate their own opinions about and answers to the complex issues of the present day.

INTRODUCTION

When newspaper headlines informed the world on February 22, 1997, that a sheep (whimsically named Dolly, after country-western singer Dolly Parton) had been cloned from an udder cell of an adult ewe in Scotland, many people panicked. If a mammal had been cloned, they assumed, the cloning of humans could not be far behind.

Scenarios of what human cloning might mean, most of them nightmarish, almost immediately began to appear in opinion articles and television commentaries. Indeed, fears of human cloning were so strong that less than a week after the sheep cloning announcement, U.S. President Bill Clinton called for a ban on the use of federal funding for any research that might lead to human cloning, and bills intended to ban all human cloning research outright were quickly introduced in both the Senate and the House of Representatives. Among ordinary citizens, Jean Bethke Elshtain reports in the *New Republic*, "For every wild-eyed optimist [about human cloning] there are, by my count, a good half-dozen alarmists."

These fears had many origins. Some were deeply embedded in cultural and religious beliefs. Others, however, may well have arisen in part from watching too many movies. Films about human cloning have appeared from time to time at least since the 1970s, and most have been negative. They have echoed concerns expressed—and to some extent shaped—by earlier novels that have portrayed renegade scientists who challenge nature only to reap terrible results, such as Mary Shelley's nineteenth-century tale *Frankenstein*, or societies that use technology to control their members, such as Aldous Huxley's 1932 novel *Brave New World*.

Movies about cloning have sometimes made for compelling viewing, but their scientific accuracy has typically been questionable. Take, for example, the 1978 movie *Boys from Brazil*. In this film, based on a novel by Ira Levin, Nazi doctor Josef Mengele flees to Brazil after the collapse of the Nazi government and there attempts to use tissue from Adolf Hitler's body to clone an army of new Hitlers. This idea is similar to a concern expressed in many opinion pieces about human cloning. Sandy Grady, for example, writes in the *Philadelphia Daily News*, "Will we be able to order biological twins of ourselves like take-out pizzas? Could we clone a basketball team of Michael Jordans? Could we tweak some DNA and reproduce Einstein? Worse, could Saddam Hussein make 10 copies of himself in some underground Baghdad lab?" Some commentators also worry that people who clone themselves will expect the results to be carbon copies rather than distinct individuals. Protestant ethicist Allen Verhey of Hope College in Holland, Michigan, warns that the availability of human cloning might make parents "think of their children as products."

However, many scientists have argued that the assumption that a human clone would share the original's personality and talents is mistaken. On the contrary, most scientists agree that environment is at least as important as genes in producing personality. They point out that although identical twins—who possess exactly the same genes and are thus natural human clones—may be remarkably similar in appearance and some aspects of behavior, they are distinctly different individuals with unique personalities. *Multiplicity*, a 1996 comedy about cloning starring Michael Keaton, conveyed another common misconception about cloning: The human clones in *Multiplicity* are created as instant adults. In actuality, even clones created from adult cells take just as long to develop as any other creatures of their kind. A clone of an adult human, being born at a much later time than the original, would inevitably be raised in a different environment and would differ even more from his or her "parent" than twins differ.

This is not to say that all concerns about human cloning are flawed or unwarranted. In fact, the National Bioethics Advisory Commission, whom President Clinton asked to study the cloning issue, had significant scientific backing for their June 1997 recommendation that human cloning should be prohibited, at least for the present, because the technology is still too risky for the potential fetus. The commission pointed out, for instance, that Dolly's was the only successful birth out of 277 tries. Similarly, scientific reports released in early 1999 revealed disturbing evidence that Dolly and other adult-cell clones may age prematurely or suffer from other hidden defects. Although Dolly remains healthy as of this writing, some other animal clones have died prematurely, which is a cause of concern to many researchers.

The general public reacted to Dolly's existence with concerns about the possibility of imminent human cloning, but to the scientific world, Dolly's importance lay elsewhere. From a scientific point of view, Dolly's significance was not that she was a clone—the first experiments with animal clones (tadpoles) had taken place some thirty years before, and even mammals were being cloned by the mid-1980s. Rather, Dolly's creation was unusual in that she was the first cloned animal to be made from a mature cell taken from an adult animal's body. Before the Scottish scientists produced Dolly, biologists had believed that only cells from very early embryos could be used to clone whole animals. In later stages of fetal development, most cells become specialized (a blood cell, for instance, is physically and chemically very different from a nerve cell), and in the process they lose the power to form other cells of types different from their own. Scientists thought these changes were irreversible until Ian Wilmut and his team at the Roslin Institute proved them wrong.

The Roslin scientists found that they could reverse the maturation process by depriving a specialized cell of nutrients for several days. They applied this treatment to a mature udder cell from a six-year-old ewe, then put the cell's nucleus into an unfertilized sheep egg cell from which the nucleus had been removed. Finally, they fused the two with a jolt of electricity. This technique, called somatic cell nuclear transfer, had been used before to produce clones from embryonic cells, but Dolly represented the first time it had been used successfully with a mature cell. The fact that the changes that occur in a cell's DNA as it matures could be sufficiently reversed to allow it to develop into a whole new organism offers many research possibilities, including a better understanding of normal embryonic development and cancer (in which cells fail to mature normally). As Diana Lutz writes in *The Sciences*, this fact, not the possibility of cloning humans, was what excited scientists about Dolly's creation.

Supporters maintain that cloning has many potential benefits. For instance, Elizabeth Pennisi explains in the January 30, 1998, issue of *Science* that the goal of the scientists who created Dolly is to produce herds of cloned animals with inserted genes that will cause their bodies to make human hormones, drugs, and other substances useful in medicine. Although some animal rights activists have questioned the ethics of using animals in this way, such procedures could lower the cost of compounds currently available only in very tiny amounts. Other supporters of human cloning, such as Mark Sauer, an infertility expert at Columbia Presbyterian Medical Center in New York, claim that it could bring new hope to infertile couples who desperately want a child and are unable to have one in any other way.

Movies and novels can be very entertaining, and the best of them can be thought-provoking as well. They can help people imagine the social effects of new technologies, a subject that strictly scientific accounts often do not explore. In order to make reasoned judgments about a groundbreaking technology such as cloning, however, it is essential to try to understand the science behind the technology as well. As Harold T. Shapiro, president of Princeton University and head of the National Bioethics Advisory Commission, states, "a great deal more widespread education and deliberation" will be needed to resolve the moral and legal issues surrounding human cloning.

Scientific and media views of human cloning are just two of the subjects addressed in *Cloning: Contemporary Issues Companion*. The book also provides information and pro-and-con opinions concerning animal cloning, key scientists involved in cloning, medical research related to cloning, legal and moral issues raised by the possibility of human cloning, and the question of whether human cloning should be temporarily or permanently banned. If a cloned human child is finally created—as many commentators feel will happen in the near future, whether the technology is legally banned or not—discussion on all these issues is sure to increase.

CHAPTER 1

THE SCIENCE OF CLONING

DOLLY, THE AMAZING SHEEP CLONE

Michael Specter and Gina Kolata

As Michael Specter and Gina Kolata explain in the following selection, the world was thunderstruck when Ian Wilmut and other scientists at the Roslin Institute in Scotland announced on February 22, 1997, that they had cloned a sheep from an udder cell of a six-year-old ewe. The lamb, named Dolly after country-western singer Dolly Parton, was not the first animal or even mammal to be cloned, Specter and Kolata say, but she was the first to be made from an adult body cell, something that many scientists thought would never be accomplished. The authors describe how Wilmut and his coworkers achieved this feat, the research that led up to it, and why it is important to science and ethics. Specter writes frequently for the *New York Times* and the *New Yorker* on science and other topics. Kolata is a science reporter for the *New York Times* and a former writer for *Science* magazine. She is also the author of *The Road to Dolly and the Path Ahead*, a book about cloning research.

Charles Darwin was so terrified when he discovered that mankind had not been specially separated from all other animals by God that it took him two decades to find the courage to publish the work that forever altered the way humans look at life on Earth. Albert Einstein, so outwardly serene, once said that after the theory of relativity stormed into his mind as a young man, it never again left him, not even for a minute.

But Dr. Ian Wilmut, the 52-year-old embryologist who astonished the world on Feb. 22, 1997, by announcing that he had created the first animal cloned from an adult—a lamb named Dolly—seems almost oblivious to the profound and disquieting implications of his work. Perhaps no achievement in modern biology promises to solve more problems than the possibility of regular, successful genetic manipulation. But certainly none carries a more ominous burden of fear and misunderstanding.

"I am not a fool," Dr. Wilmut said in his cluttered lab, during a long conversation in which he reviewed the fitful 25-year odyssey that led to his electrifying accomplishment and unwanted fame. "I know what is bothering people about all this. I understand why the world is suddenly at my door. But this is my work. It has always been my work, and it doesn't have anything to do with creating copies of human beings. I am not haunted by what I do, if that is what you want to know. I sleep very well at night."

Founder of a New Era

Yet by scraping a few cells from the udder of a 6-year-old ewe, then fusing them into a specially altered egg cell from another sheep, Dr. Wilmut and his colleagues at the Roslin Institute, seven miles from Edinburgh, have suddenly pried open one of the most forbidden—and tantalizing—doors of modern life.

People have been obsessed with the possibility of building humans for centuries, even before Mary Shelley wrote *Frankenstein* in 1818. Still, so few legitimate researchers actually thought it was possible to create an identical genetic copy of an adult animal that Dr. Wilmut may well have been the only man trying to do it, a contrast with the fiery competition that has become the hallmark of modern molecular biology.

Dr. Wilmut, a meek and affable researcher who lives in a village where sheep outnumber people, grew more disheveled and harried as the pressure-filled week wore on. A $60,000-a-year government employee at the institute, Scotland's leading animal research laboratory, Dr. Wilmut does not stand to earn more than $25,000 in royalties if his breakthrough is commercially successful.

"I give everything away," he said. "I want to understand things."

Dr. Wilmut has made no conscious effort to improve on science fiction in his work; he said, in fact, that he rarely read it. A quiet man whose wife is an elder in the Church of Scotland but who says he "does not have a belief in God," Dr. Wilmut is the least sensational of scientists. Asked the inevitable questions about cloning human beings, he patiently conceded that it might now become possible but added that he would "find it repugnant."

Dr. Wilmut's objectives have always been prosaic and direct: he has spent his life trying to make livestock healthy, more efficient and better able to serve humanity. In creating Dolly, his goal—like that of many other researchers around the world—was to turn animals into factories churning out proteins that can be used as drugs. Even though the work is early and tentative, and it needs many improvements before it can be used, no scientists have stepped forward to say that they doubt its authenticity.

Many scientists say they are certain that the day will eventually come when humans can also be cloned. About a week after Wilmut's announcement, scientists in Oregon said they had cloned rhesus monkeys from very early embryo cells. That is not the same as cloning the more sophisticated cells of an adult animal, or even a developing fetus. But any kind of cloning in primates brings the work closer to human beings. That is why what has happened in Roslin has rapidly begun to resonate far beyond the tufted glens and heather hills of Scotland. In much the way that the Wright Brothers at Kitty Hawk freed humanity of a restriction once considered eternal, human existence suddenly seems to have taken on a dramatic new dimension.

The eventual impact of this particular experiment on business and science may not be known for years. But it will almost certainly cast important new light on basic biological science.

Already, even the simplest questions about the creation of Dolly provoke answers that demonstrate how profound and novel the research has been. Asked if the lamb should be considered 7 months old, which is how long she has been alive, or 6 years old, since it is a genetic replica of a 6-year-old sheep, Dr. Wilmut's clear blue eyes clouded for a moment. "I can't answer that," he said. "We just don't know. There are many things here we will have to find out."

Years of Labor in the Barnyard

The Scots have an old tongue twister of an adage that says "many a mickle make a muckle," or, little things add up to big things. It is certainly true of the cloning of Dolly, who had her conceptual birth in a conversation in an Irish bar more than a decade ago and who was born after a series of painstaking experiments, years of doubt and several final all-night vigils—one bleating little lamb among nearly 300 abject failures.

While the world has become transfixed by the idea of creating identical copies from frozen cells, that was not the result that Dr. Wilmut, or any other scientists interviewed for this article, considers the most significant part of the research.

The true object of those years of labor was to find better ways to alter the genetic makeup of farm animals to create herds capable of providing better food or any chemical a consumer might want. In theory, genes could be altered so animals would produce better meat, eggs, wool or milk. Animals could be made more resistant to disease. Researchers even talk about breeding cows that could deliver low-fat milk straight from the udder.

"The overall aim is actually not, primarily, to make copies," Dr. Wilmut said, interrupted constantly by the institute's feed mill as it noisily blew off steam. "It's to make precise genetic changes in cells."

Obscure as he may seem to those outside his field, Ian Wilmut has been quietly pushing the borders of reproductive science for decades. In 1973, having just completed his doctorate at Cambridge, he produced the first calf born from a frozen embryo. Cows give birth to no more than 5 or 10 calves in a lifetime. By taking frozen embryos produced by cows that provide the best meat and milk, thawing them and transferring them to surrogate mothers, Dr. Wilmut enabled cattle breeders to increase the quality of their herds immensely.

Since then, while always harboring at least some doubt that cloning was really possible, he has struggled to isolate and transfer genetic traits that would improve the utility of farm animals.

A Pivotal Rumor

In 1986, while in Ireland for a scientific meeting, Dr. Wilmut heard something during a casual conversation in a bar that caught his attention and convinced him that cloning large farm animals was indeed possible. "It was just a bar-time story," he recalled, in the slight brogue he has acquired after living here for 25 years. "Not even straight from the horse's mouth."

What he heard was the rumor—true, it turned out—that another scientist had created a lamb clone from an already developing embryo. It was enough to push him in a direction that had already been abandoned by most of his colleagues.

By the early 1980s, many researchers had grown discouraged about the practicalities of cloning because of a hurdle that had come to seem insurmountable. Every cell in the body originates from a single fertilized egg, which contains in its DNA all the information needed to construct a whole organism. That fertilized egg cell grows and divides. The new cells slowly take on special properties, developing into skin, or blood or bones, for example. But each cell, however specialized, still carries in its nucleus a full complement of DNA, a complete blueprint for an organism.

The problem for scientists was stark and unavoidable: It was assumed that the nucleus of a mature cell, which has developed, or differentiated, so it could carry out a specific function in the body, simply could not be made to function like the nucleus of an embryo that had yet to begin the process of learning to play its special role. Even though the DNA, with all the necessary genes, was in the differentiated cell, the issue was how to turn it on so it would direct the process of growth that begins with the egg. The essential question for cloning researchers was whether the genes in an adult cell could still be used to create a new animal with the same genes.

The pivotal rumor Dr. Wilmut heard at the meeting in Ireland was that a Danish embryologist, Dr. Steen M. Willadsen, then working at Grenada Genetics in Texas, had managed to clone a sheep using a cell from an embryo that was already developing.

The story, which came from a veterinarian named Geoff Mahon, who worked at the same company, went beyond the research that Dr. Willadsen would publish later that year on cloning sheep from early embryos. Dr. Willadsen said in a telephone interview from his home in Florida that he had indeed done the more advanced work but had never published it.

What he did publish was the result of successfully cloning sheep from very early embryo cells: the first cloning of a mammal. Dr. Willadsen tried that experiment with three sheep eggs. In each case, he removed the egg's nucleus, with all its genetic information, and fused that egg, now bereft of instructions on how to grow, with a cell from a growing embryo. If the egg could use the other cell's genetic information to grow itself into a lamb, the experiment would be a success. It worked.

"The reality is that the very first experiment I did, which involved only three eggs, was successful," Dr. Willadsen said. "It gave me two lambs. They were dead on arrival, but the next one we got was alive." The paper was eventually published in *Nature*, the influential British science journal that in February 1997 published Dr. Wilmut's news of Dolly, and it created a sensation.

But it was the rumor of the unpublished work that captivated Dr. Wilmut. If it was possible to clone using an already differentiated embryonic cell, it was time to take another look at cloning an adult, Dr. Wilmut decided. "I thought if that story was true—and remember, it was just a bar-time story—if it was true, we could get those cells from farm animals," he said. And, he thought, he might even be able to make copies of animals from more mature embryos or eventually from an adult.

When Dr. Wilmut flew back to Scotland, he was already dreaming of Dolly. When he was flying back over the Irish Sea with a colleague, he said, "we were already making plans to try to get funds to start this work."

From Daydreams to Successes

Dr. Wilmut's dominance of the field grew from that day, almost by default. He was nearly alone, out on a limb. His tumultuous field seemed to have run out of steam. Many of its leaders and its students had departed, going to medical school and becoming doctors or accepting lucrative positions at in vitro fertilization centers, helping infertile couples have babies. Two of its stars published a famous paper concluding that cloning an adult animal was impossible, dashing cold water on their eager colleagues. Companies, formed in a flush of enthusiasm a decade earlier, folded by the early 1990s.

Most of the few cloning researchers left were focused on a much easier task. They were cloning cells from early embryos that had not yet specialized. And even though some had achieved stunning successes, none were about to try cloning an adult or even cells from mature embryos. It just did not seem possible.

The idea of cloning had tantalized scientists since 1938. When no one even knew what genetic material consisted of, the first modern embryologist, Dr. Hans Spemann of Germany, proposed what he called a "fantastical experiment": taking the nucleus out of an egg cell and replacing it with a nucleus from another cell. In short, he suggested that scientists try to clone.

But no one could do it, said Dr. Randall S. Prather, a cloning researcher at the University of Missouri in Columbia, because the technology was not advanced enough. It would be another 14 years before anyone could try to clone, and then they did it with frogs, whose eggs are enormous compared with those of mammals, making them far easier to manipulate. Dr. Spemann, who died in 1941, never saw his idea carried to fruition.

In fact, frogs were not successfully cloned until the 1970s. The work was done by Dr. John Gurdon, who now teaches at Cambridge University. Even though the frogs never reached adulthood, the technique used was a milestone. He replaced the nucleus of a frog egg, one large cell, with that of another cell from another frog.

It was the beginning of nuclear transfer experiments, which had the goal of getting the newly transplanted genes to direct the development of the embryo. But the frog studies seemed to indicate that cloning could go only so far. Although scientists could transfer nuclei from adult cells to egg cells, the frogs only developed to tadpoles, and they always died.

Most researchers at the time thought even that sort of limited cloning success depended on something special about frogs. "For years, it was thought that you could never do that in mammals," said Dr. Neal First of the University of Wisconsin, who has been Dr. Wilmut's most devoted competitor.

Dashed Hopes

In 1981, after some rapid advances in technology, two investigators published a paper that galvanized the world. It seemed to say that mammals could be cloned—at least from embryo cells. But, in a crushing blow to those in the field, the research turned [out] to be a fraud.

The investigators, Dr. Karl Illmensee of the University of Geneva and Dr. Peter Hoppe of the Jackson Laboratory in Bar Harbor, Me., claimed that they had transplanted the nuclei of mouse embryo cells into mouse eggs and produced three live mice that were clones of the embryos. Their mice were on the cover of the prestigious journal *Science*, and their work created a sensation.

"Everyone thought that article was right," said Dr. Brigid Hogan, a mouse embryologist at Vanderbilt University in Nashville. Dr. Illmensee, the senior author, "was getting enormous publicity and exposure, and accolades," Dr. Hogan said.

Two years later, however, two other scientists, Dr. James McGrath and Dr. Davor Solter, working at the Wistar Institute in Philadelphia, reported in *Science* that they could not repeat the mouse experiment. They concluded their paper with the disheartening statement that the "cloning of mammals by simple nuclear transfer is impossible."

After a lengthy inquiry, it was discovered that Dr. Illmensee had faked his results.

Leaders in the field were shattered. Dr. McGrath gave up cloning, got an M.D. degree and is now a genetics professor at Yale University. Dr. Solter gave up cloning and is now the director of the Max Planck Institute in Freiburg, Germany. Most research centers abandoned the work completely.

"Man, it was depressing," said Dr. James M. Robl, a cloning researcher at the University of Massachusetts in Amherst. After the paper by Dr. Illmensee, "we all thought we would be cloning animals like crazy," Dr. Robl said. He had pursued research to try to clone cows and pigs. Suddenly, it seemed as though he was wasting his time.

"We had a famous scientist come through the lab," Dr. Robl said. "I showed him with all enthusiasm all the work I was doing. He looked at me with a very serious look on his face and said, 'Why are you doing this?'"

Making Cells Hibernate

But not everyone was despondent. A few investigators forged on. One of them was Dr. Keith Campbell, a charismatic 42-year-old biologist at the Roslin Institute who specializes in studying the life cycle of the cell. Dr. Campbell, who joined the institute in 1991, said in an interview, "I always believed that if you could do this in a frog, you could do it in mammals." Dr. Campbell, who said he had enjoyed the cloning fantasy *The Boys from Brazil*, responded to questions about his earlier work on cloning in an interview the summer of 1996, saying "We're only accelerating what breeders have been doing for years."

Soon he had convinced his colleagues at the institute to try the experiments that eventually led to their success with Dolly. "But at that point, we still had much to learn," he said.

The most important step would be to find a way to grow clones from cells that had already developed beyond the very earliest embryonic stage. Whenever cloning had been tried with more specialized cells in the past, it had ended in failure. Until Dolly was born, nobody could be sure whether those failures were because older cells have switched off some of their genes for good or because nobody knew how to make them work properly in an egg.

Because no one knew whether cloning was even possible, it was hard to speculate about what the hurdles might be. But Dr. Campbell had what turned out to be the crucial insight. It could be, he realized, that an egg will not take up and use the genetic material from an adult cell because the cell cycles of the egg and the adult cell might be out of synchrony. All cells go through cycles in which they grow and divide, making a whole new set of chromosomes each time. In cloning, Dr. Campbell speculated, the problem might be that the egg was in one stage of its cycle while the adult cell was in another.

Dr. Campbell decided that rather than try to catch a cell at just the right moment, perhaps he could just slow down cellular activity, nearly stopping it. Then the cell might rest in just the state he wanted so it could join with an egg.

"It dawned on me that this could be a beneficial way of utilizing the cell cycle," he said, in what may turn out to be one of scientific history's great understatements.

What he decided to do was to force the donor cells into a sort of hibernating state, by starving them of some nutrients.

In Wisconsin, Dr. First had actually beaten the Scottish group to cloning a mammal from cells from an early embryo; that occurred when a staff member in the laboratory forgot to provide the nourishing serum, inadvertently starving the cells. The result, in 1994, was four calves. But even Dr. First and his colleagues did not realize the significance of how the animals had been created.

Two years later, Drs. Wilmut and Campbell tried the starvation technique on embryo cells to produce Megan and Morag, the world's first cloned sheep and, until now, the most famous sheep in history. Their creation really laid the foundation for what happened with Dolly, for Dr. Campbell succeeded in doing an end run around the problem of coordinating the cycles of the donor cell with the recipient egg.

Today, Megan and Morag munch contentedly in the same straw-covered pen with the new star of the Roslin Institute, angelic little Dolly. Megan and Morag seem completely normal, if slightly spoiled.

Megan is now expecting, and she got pregnant the old-fashioned way. "It will always be the preferred way of having children," Dr. Campbell said jokingly. "Why would anyone want to clone, anyway? It's far too expensive and a lot less fun than the original method."

A Star Is Born

When the scientists moved on to cloning a fully grown sheep, they decided to use udder, or mammary, cells, and that is how Dolly got her name. She was named after the country singer Dolly Parton, whose mammary cells, Dr. Wilmut said, are equally famous.

In the experiment that produced Dolly, Dr. Wilmut's team removed cells from the udder of a 6-year-old sheep. The cells were then preserved in test tubes so the investigators would have genetic material to use in DNA fingerprinting—required to prove that Dolly was indeed a clone. In fact, by the time Dolly was born, her progenitor had died.

The trick, Dr. Wilmut said, was the starvation of the adult cells. "You greatly reduce serum concentration for five days," Dr. Wilmut said. "That's the novel approach. That's what we submitted a patent for." And that is why the team was silent about the lamb's birth for months. Until the patent was applied for, nobody wanted the news to spread.

But success is a relative concept. Even Dr. Campbell's technique has failed far more often than it has succeeded. Dolly was the only lamb to survive from 277 eggs that had been fused with adult cells. Nobody knows, or can know, until the work is repeated, whether the researchers were lucky to get one lamb—whether in fact that one lamb was one in a million and not just one in 277 or whether the scientists will become more proficient with more refinement.

The cell fusion that produced Dolly was done in the last week of January 1996. When the resulting embryo reached the six-day stage, it was implanted in a ewe. Dolly's existence as a growing fetus was first discovered on March 20, the 48th day of her surrogate mother's pregnancy. After that, the ewe was scanned with ultrasound, first each month and then, as interest grew, every two weeks.

"Every time you scanned, you were always hoping you were going to get a heartbeat and a live fetus," said John Bracken, the researcher who monitored the pregnancy.

"You could see the head structure, the movement of the legs, the ribs," he said. "And when you actually identified a heart that was beating, there was a great sense of relief and satisfaction. It was as normal a pregnancy as you could have."

On July 5 at 4 P.M., Dolly was born in a shed down the road from the institute. Mr. Bracken, a few members of the farm staff and the local veterinarian attended. It was a normal birth, head and forelegs first. She weighed 6.6 kilograms, about 14½ pounds, and she was healthy.

Because it was summer, the few staff members present were very busy. There was no celebration.

"We phoned up the road to inform Ian Wilmut and Dr. Campbell," Mr. Bracken said.

But Dr. Wilmut does not remember the call. He does not even remember when he heard about Dolly's birth.

"I even asked my wife if she could recall me coming home doing cartwheels down the corridor, and she could not," he said.

COULD DOLLY BE A HOAX?

Michael Waldholz

In the following selection, Michael Waldholz, a Pulitzer Prize–winning reporter for the *Wall Street Journal*, considers whether the amazing announcement that Ian Wilmut and his coworkers had cloned a lamb, Dolly, from an adult body cell could be a hoax. He reviews previous scientific errors and hoaxes, such as "cold fusion" and Piltdown Man, but he also offers evidence that the achievement claimed by Wilmut and his colleagues is real. In addition, he describes ways that scientists may build on Wilmut's groundbreaking research. Waldholz is the author of the book *Curing Cancer: The Story of the Men and Women Unlocking the Secrets of Our Deadliest Illness*.

The cloning of an adult mammal—a sheep named Dolly—came as such a shock that it has some people wondering if the Scottish researchers who delivered the lamb were pulling the wool over our eyes.

Because the breakthrough had been assumed to be undoable, skeptics may well ask: Why did the scientists wait seven months to deliver Dolly's birth announcement? Why is her DNA "mother" no longer alive? And who are these guys, anyway?

The fact is that—as with any scientific advance—until the feat can be duplicated elsewhere, the Scottish team's claims will have to be taken on faith. "I know and trust them, but if what they did isn't repeatable, they'll pay for it with their careers," says Randall Prather, a veterinary scientist at the University of Missouri at Columbia.

Even *Nature*, the prestigious journal that published the new cloning research in February 1997, doesn't demand physical evidence. Instead, the journal, whose reports are reviewed before publication by leading experts in the field, relies on the veracity of the scientists who are describing their work. "I reviewed the paper and it convinced me," says one of the *Nature* reviewers, Colin Stewart, a researcher at the National Cancer Institute lab in Frederick, Md.

Genetics experts say there are plenty of good reasons to believe. . . . Perhaps the most compelling argument in favor of the veracity of the breakthrough is the embryologist involved, Ian Wilmut of the Roslin

Institute in Edinburgh. Unknown to the general public before the news of the cloning was published, he is a giant in the field of embryo science.

"If it were anyone else I'd be very suspicious," says Curtis Youngs, a professor of animal science at Iowa State University. "But Ian is brilliant, a man of tremendous integrity. I have absolute trust that what he says is true."

In the early 1970s, Dr. Wilmut pioneered the techniques for freezing and preserving embryos, which revolutionized veterinary science and contributed to in vitro fertilization research. In recent years he has produced a series of breakthroughs that provided a rational foundation for the next step—the creation of Dolly.

History of Hoaxes

Nonetheless, the history of science is rife with sensational claims that later proved to be flops or flat-out hoaxes. "Most remarkable discoveries turn out to be wrong, and this one was very remarkable," says author Gary Taubes, who has written extensively on scientific lapses and frauds. "Science is a human endeavor. People are going to cheat" and make mistakes.

In 1989 two Utah-based scientists stunned the world by announcing they had achieved nuclear fusion—a cataclysmic reaction that normally occurs on the surface of the sun—in a cup of cold water. "Cold fusion" promised a cheap, unlimited source of energy, but physicists soon said it was impossible. As of 1997, despite prodigious efforts, no one has been able to conclusively duplicate the experiment. The Utah researchers "convinced themselves they were right," says Mr. Taubes, who wrote a book about the incident.

Some outright hoaxes hold up for decades. In 1912, an amateur archeologist said he had discovered, in a gravel pit in Piltdown, England, an ancient human skull fragment that seemed to be a spectacular confirmation of the theory that humans' large brains were the first feature to differentiate humans from apes. But in the 1950s scientists proved that "Piltdown man" was a prank. It took another 40 years to identify the hoaxer as a museum curator who had stained human bones to make them appear ancient.

Even in the pursuit of clones, at least one milestone turned out to be an illusion. In 1982 Karl Illmensee at the University of Geneva, collaborating with a scientist at Jackson Laboratory in Maine, reported success in cloning mouse embryos. Soon after, however, several lab mates reported that Dr. Illmensee wasn't even present when he claimed to have conducted some of the key experiments. Several investigations found no evidence of fraud but noted sufficient errors and irregularities in data collection to cast doubt on the results.

Since then, however, cloning has made huge progress, which is why embryo experts say the work in Scotland is credible. It builds on a cascade of experiments over the past decade. In recent years scientists have been able to replace the nucleus in one embryo with that of another; they have also successfully cloned embryos whose genes have been manipulated. In 1996 Dr. Wilmut pulled off the transfer of a nucleus from a developing embryo into an unfertilized egg, the first time that had been done using an egg rather than an embryo.

Using an adult cell rather than an embryo is far more difficult, though. To succeed requires overthrowing "the accepted dogma of developmental biology," says Robert Wall, a geneticist at the U.S. Department of Agriculture in Beltsville, Md. The dogma: Once the genes in a cell have matured enough to specialize and form a specific tissue such as liver cells or nerve cells, they can't be reprogrammed to act like unspecialized, embryonic genes open to new instructions.

"When Ian told us that he was going to transfer the [genes] from a [mature] cell" into an unfertilized egg, "we all rolled our eyes," says Jorge Piedrahita, a researcher at Texas A&M University.

In the experiment, Dr. Wilmut's group extracted the DNA in the nucleus of a cell from the udder of a six-year-old ewe. The cells were preserved in a test tube for many subsequent experiments. By the time Dolly was cloned, her "mom" had died. The "adult" mammary genes were eventually put to sleep, essentially, by being deprived of nutrients.

This idle state let them act more like embryonic DNA once they were inserted in place of the DNA in an unfertilized egg cell. To the amazement of many in the field, the recipient egg cell roused the mature cell's genes and stimulated them to split and grow as if they were inside a normal embryo. Dr. Wilmut withheld news of his success because his research is funded by a company, PPL Therapeutics PLC, that is seeking a patent for the new technology.

The breakthrough—coaxing adult, specialized DNA to return to its roots and act like undeveloped cells all over again—"is his great contribution," Dr. Piedrahita says. "Ian Wilmut will be renowned centuries from now not for cloning a mammal but for proving that a nucleus from a [mature] cell can be made to behave like the nucleus of an embryo."

Building on a Breakthrough

Now, "we're going to build on Ian's discovery," says James Robl, a researcher at the University of Massachusetts. Dr. Robl says his team soon hopes to produce an offspring with altered genetic material that was originally derived from a mature animal's cell. Eventually the technology could help produce heartier breeds of livestock, animals whose organs can be used for human transplantation, and cows and sheep whose milk contains important human hormones and drugs. "I would expect Ian is already trying to do this, too," Dr. Robl says. "He's opened a whole new avenue of research that is going to get very exciting very soon."

Neal First, an animal researcher at the University of Wisconsin, says he heard rumors of Dr. Wilmut's cloning experiment in the autumn of 1996. "But I said I'd believe it when I saw it," he says. "Now I've seen [the research report] and I believe it." Dr. First plans to use the new technique to try to one-up Dr. Wilmut: He plans to clone a cow.

THE REAL IMPORTANCE OF DOLLY

Diana Lutz

In their tendency to see the cloned sheep, Dolly, as a kind of Frankenstein's monster, the public missed her true importance to science, Diana Lutz claims in the following essay. Lutz describes a press conference held by the New York Academy of Sciences in New York City a few days after Dolly's existence was announced on February 22, 1997. During this conference, she explains, several scientists tried to put Dolly's creation into an accurate scientific and ethical context. While media attention largely focused on the possibility of human cloning that Dolly's creation raised, she reports, the scientists argued that Dolly's existence was unlikely to lead to research into human cloning. According to Lutz, the scientists maintained that Dolly's real value is that she shows that modifications in a cell's DNA that occur as the cell matures and develops into a particular type can be reversed, making the cell capable of developing into a whole organism. Lutz is a frequent book reviewer for *Horn Book*, a magazine devoted to education.

"Frankenstein's monster" cried the popular press in response to the announcement from Scotland that an adult ewe had been cloned in February 1997. The reference was probably inevitable: the mammary-gland cell that supplied the DNA for making a lamb was an "adult body part," fused with an egg cell by a jolt of electric current. But the media clearly didn't have in mind the pitiful, self-loathing student of Goethe in Mary Wollstonecraft Shelley's classic novel. What they meant instead was Boris Karloff with bolts through his neck. Does it matter?

Banning the Monster

At a background press conference on cloning and its ethical implications, convened at the New York Academy of Sciences in New York City a few days after the epochal announcement was made, an eight-member panel—a developmental biologist, three medical doctors, a biotechnology company executive, a lawyer, a bioethicist and a

Excerpted from "Hello, Hello, Dolly, Dolly," by Diana Lutz, *The Sciences*, May/June 1997. Reprinted by permission of *The Sciences*. Individual subscriptions are $28 per year. Write to *The Sciences*, 2 East 63rd St., New York, NY 10021.

rabbi—agreed that the cloning story had turned into a caricature of itself. In the process much of the scientific news had been missed.

Alan R. Fleischman, senior vice president for medical and academic affairs at the New York Academy of Medicine, called the visceral reaction of the public to the announcement more interesting than the scientific achievement itself. Erik Parens, a bioethicist at the Hastings Center in Briarcliff Manor, New York, agreed. Both cheerleaders and doomsayers alike share the same mistaken idea, he said: that you can clone a self. "But as everyone in this room knows, you can't clone a self," Parens noted, "because a self is a function of infinitely more than one's genetic material."

Several panelists argued that the issue of cloning the human species is a red herring, simply because it isn't practical, desirable or even likely. Parens, building on his point about the cloning of a self, wondered what effect that idea, however mistaken it might be, would have on culture and politics.

Roy M. Goodman, a New York state senator who was in the audience at the press conference, gave Parens what amounted to an instant answer to the political question. What did the panelists think, he asked, of legislation proposed by New York State Senator John J. Marchi that would make it a felony to proceed with human cloning in the state? Fleischman said he saw no reason for rapid-fire legislation, because the ability to clone a human won't come any time soon. Marie A. DiBerardino, a developmental biologist at the Medical College of Pennsylvania–Hahnemann School of Medicine in Philadelphia, agreed, though she said she would welcome a ban on human cloning.

Why Dolly Is Important

Having disposed of Frankenstein's monster, the panelists sought to give a better account of just what the embryologist Ian Wilmut and his coworkers at the Roslin Institute in Edinburgh, Scotland, had done in their now-famous experiment. According to DiBerardino, who characterized herself as "the oldest cloner, or nearly the oldest one," the procedure of nuclear transplantation used by Wilmut was developed by Robert Briggs and Thomas J. King in Philadelphia as long ago as 1952. First tested in amphibians, the procedure was soon extended to insects and fishes and, in the 1980s, to mammals.

All those experiments, however, began with embryonic cells—cells that have not yet differentiated, or become specialized for a particular physiological function. Although Wilmut admitted he did not determine whether the donor DNA came from mammary tissue or from a relatively undifferentiated mammary stem cell, the donor cell was still far more specialized than any cell from an embryo. When Wilmut's new lamb, Dolly, emerged from the transferred DNA of an adult sheep, what Wilmut had achieved was, most of all, the cloning of a differentiated cell.

That was partly what made the popular accounts of the sheep cloning so puzzling—and so misguided. Many of them distinguished between cloning an embryo and cloning an adult. But had the point really sunk in, the story of cloning a monkey from an embryo cell, which broke soon after Wilmut's announcement, would have been ignored.

Clones Are Not Duplicates

The panelists also sought to correct another widespread misconception: that an adult clone is a carbon copy of its parent. The clone and parent need not be identical for at least three reasons. First, not all of a cell's DNA lies in its nucleus. A cell's mitochondria, the organelles that fuel cellular reactions, also contain DNA, and the mitochondrial DNA invariably comes from the egg. Unless the nuclear DNA from the parent were transplanted into an egg from the parent's mother, a clone would not genetically match the parent at every level. Second, the DNA in an adult cell differs from the DNA in a fetal cell by a lifetime's worth of damage.

But the main point, stressed by several panelists besides Parens, is that even "identical" twins are not identical. Any animal is the product not just of its genetics but also of its environment, both in utero and after birth.

Not everyone agreed that such distinctions are important. Rabbi Marc Gellman of Temple Beth Torah in Melville, New York, rebuked his fellow panel members for logic chopping. "There is a wisdom in common people, in ordinary people," he said, "who have been unfairly demeaned by people who view their lack of knowledge about haploid and diploid and totipotent embryos as somehow a disqualification to have moral sensibilities. . . . There is a strong and real understanding that we are not our own creators. And this technology undermines that fundamental belief in the most powerful and disturbing way possible."

The panelists, however, seemed less arrogant than dismayed— deeply dismayed that much of the public discussion assumes that people are the products of their DNA alone. The irony of the New York Academy session was that men and women who have devoted their lives to the study of DNA were urging the audience not to overestimate its role in their lives.

Reversing Changes in DNA

So if cloning human beings is impractical, unethical and unlikely, what use is Wilmut's experiment? Dean L. Engelhardt, senior vice president of Enzo Biochem, Inc., in New York City, said cloning would improve the process of animal husbandry by enabling animals with genetically desirable traits to be bred more efficiently. Indeed, the more sober newspaper accounts suggested that Wilmut's work probably would lead to the cloning of the nation's best milk cows rather than of humans.

But Jon W. Gordon, a professor in the departments of obstetrics/ gynecology and geriatrics at Mount Sinai School of Medicine in New York City, raised practical objections to such ideas. Gordon has done in-vitro fertilization with humans as well as mice, and he has done gene-transfer experiments in mice. Cloning will never replace selective breeding as a method of genetic engineering in animals, he said. For one thing, cloning halts any further progress: once you've cloned a cow, for instance, you can't keep breeding that version for greater milk production. Gordon also opposes schemes for using cloning to mass-produce "therapeutic" tissues or proteins.

If cloning is not a panacea for agricultural or medical problems, why does it interest scientists? DiBerardino may have given the best answer when she said: "It's possible to take a kidney carcinoma cell nucleus from a frog and inject it into an enucleated egg and get a swimming tadpole. Not more tumor masses, but a swimming tadpole. So the molecular milieu of the [egg] cytoplasm has dictated that the normal genes turn on and the genes causing the proliferation turn off." In other words, every cell holds within its nucleus the information needed to make every kind of cell in the body and, indeed, to make a whole frog, or rabbit, or person. But during development those nuclear recipes are somehow selectively turned off or blocked. As a result, the cell is no longer totipotent, or capable of giving rise to an entire, fertile organism. The lesson of Wilmut's experiment is that the modification of the nuclear DNA is reversible.

CLONING MONKEYS: ARE HUMANS NEXT?

Virginia Morell

A few days after the cloning of Dolly the sheep by scientists in Scotland was made public in February 1997, researchers in Oregon announced that they had cloned two rhesus monkeys. Although the monkeys were cloned from embryonic cells, a less unusual achievement than Dolly's creation from an adult cell, Virginia Morell explains that the rhesus monkey research is important because it involves animals much more closely related to humans than sheep are. Morell describes the monkey research in detail and compares the cloning technique used with the type employed to create Dolly. She also considers the implications of the research for the possibility of human cloning. Morell is a contributing editor for *Discover* magazine and the author of the book *Ancestral Passions: The Leakey Family and the Quest for Humankind's Beginnings*.

In January 1998, scientists from a Massachusetts firm, Advanced Cell Technology, showed off three cloned calves, Charlie, George, and Albert, which were apparently produced via a more sophisticated (and patentable) technique than the one used to produce Dolly, the cloned sheep, in February 1997. At human fertility clinics, researchers are pursuing studies of human eggs that could lay the groundwork for cloning, although that is not the purported intent. And the National Institutes of Health (NIH) has funded two projects to clone rhesus monkeys, although only embryonic and fetal cells, not those from adults, will be used. Still, these types of studies bring human cloning closer to reality.

Good old-fashioned curiosity is pushing the field as well. "Ethics aside, I have to say as a scientist I find the technological problems fascinating," says David Ledbetter, a human geneticist at the University of Chicago, voicing a sentiment others in human reproductive biology share. "Why is this difficult to do? What will it take to make it work? How do you make a clone?"

As Ledbetter's queries suggest, making a human clone is not simply a matter of following a recipe. The journal article announcing Dolly's birth didn't spell out a formula for cloning mammals; in fact, it didn't identify the actual cell that supplied Dolly's genetic material. Yet even without that key piece of information, Dolly's appearance was utterly astounding, since most biologists believed that it was impossible to produce a cloned mammal using any adult cell. "That's what everyone thought," says Don Wolf, a senior scientist at the Oregon Regional Primate Research Center in Beaverton, who's overseeing the rhesus monkey cloning project. "But Ian Wilmut [the Scottish scientist who led the Dolly project] came up with a clever innovation, a neat trick that proved us all wrong."

Before Dolly, researchers thought that adult cells could not be induced to produce a clone because they are already differentiated. As a fertilized egg develops into an adult, it divides into two, then four, then eight identical cells. Soon, however, the cells begin to specialize, becoming bone or skin, nerve or tissue. These differentiated cells all share the same DNA—the blueprint of the body—but they follow different parts of the instructions it contains. "In a sense, they're programmed," says Wolf, and as they age, it becomes more and more difficult to reprogram them, to make them switch functions. That's exactly what the Scottish team did when they produced Dolly: they took the genetic material from a differentiated adult cell and made it behave like the genetic material in a newly fertilized egg. Their success, however, does not mean that it is now easy to reprogram a human adult cell. If anything, notes Wolf, researchers suspect that every species is unique in its requirements for setting its cellular clock back to zero.

Neti and Ditto

Low-key and soft-spoken, Wolf stepped into the cloning spotlight in 1997, when the primate center announced that he had produced two monkeys, called Neti (an acronym for "nuclear embryo transfer infant") and Ditto, using a technique similar to the one used to make Dolly. Despite Ditto's name and stories in the press, the monkeys are not identical copies of each other; they are only brother and sister. They were cloned using cells taken from two different embryos that shared an egg donor and a sperm donor. Still, their existence demonstrates that the formerly unthinkable is doable—and with primates.

Further, Wolf suspects he could produce clones from adult monkey cells as well, although, he is quick to add, he's not attempting to do so. "I have no desire to compete with the Richard Seeds [a physicist turned biologist who claimed in early 1998 that he would soon clone humans] of this world," says Wolf. "Nor do I want to see a knee-jerk reaction from our legislators that bans everything we're trying to do, particularly with techniques that have such tremendous potential for biomedical research." Already Congress has made research on human embryos off-limits to anyone receiving federal grants. Those who do not comply will have their labs shut down. It's safer, Wolf and others say, simply to avoid the subject.

A New Lab for Monkey Cloning

Wolf retired from the primate center in 1996; he came back only after receiving the two NIH grants to produce a series of cloned monkeys for medical research and is setting up his new lab in early 1998. When complete, it will occupy three rooms in one of the center's squat beige buildings. In one, two researchers dressed in lab coats are peering through their microscopes at petri dishes filled with pinkish masses of monkey tissue. Somewhere in the gelatinous mix are the eggs, each about five-thousandths of an inch across. The researchers' task is to pluck out the good ones gently with a thin glass tube called a micropipette, then place these in a fresh dish for later use. Judging from the back and neck stretches the duo indulge in during a break, their efforts require almost as much concentration as trying to induce a spoon to bend. "This is going to be the main place of activity, Room 003," says Wolf, pausing briefly to check on his group's progress, then leading the way outside, where tall pines and firs tower overhead.

Wolf moves through the lushly landscaped grounds to a nondescript conference room, where he pulls up a chair and begins explaining the enormous boon genetically identical monkeys will be to medical researchers trying, for example, to develop an AIDS vaccine. "They'd be an ideal model system," says Wolf, since they'd have identical immune systems, eliminating an important potential cause of confusion when scientists test such a vaccine or other treatment.

The center already raises rhesus monkeys for medical research; most are used in experiments here and some are sold to other medical research institutions. While awaiting their fate, the monkeys live in grassy two-acre enclosures where they pick at the grass, climb tree stumps, play, and mate, keeping an eye out for their feeders. From a distance and to the uninitiated, they all look so much alike they could easily be clones. When mature, Neti and Ditto will join one of these troops. For the time being, they're kept with other young monkeys in a smaller yet roomy cage, although no one seems sure which cage they're in or if they're even in the same one. Since their brief moment of celebrity, they've been treated like any other adolescent monkeys at the center, and since they apparently look like the other adolescents, they are no longer singled out for show.

A Cloning Recipe

To make Neti and Ditto, Wolf followed a procedure that has frequently been used in the cattle industry for producing prized breeds. It is not, however, easily done; even in cattle, only 1 to 4 percent of such pregnancies yield offspring. In light of that low percentage, Wolf's first efforts represented "a tremendous success," says Dee Schramm, a reproductive physiologist at the Wisconsin Regional Primate Research Center in Madison. From 52 transplanted embryos, Wolf produced two healthy monkeys. "Yes, that's encouraging," Wolf acknowledges, "but you really can't draw any conclusions or expectations from what we did. After all, we've only done it twice." Still, the success has encouraged Schramm and his colleague David Watkins to attempt to clone rhesus monkeys themselves. They, too, have an NIH grant for work using fetal and embryonic cells. They also plan to produce monkey embryos from adult cells, to study the differences in embryonic development among clones produced from different sorts of cells. However, they won't implant embryos produced from adults into female monkeys; only clones made from fetal and embryonic cells will be carried to term. Schramm says he hopes to have "several pairs of identical monkeys over the next two years." The work is so slow and tedious, he adds, that "I don't foresee any monkey cloning factories."

That's because it's tricky to reprogram any differentiated cell, whether embryonic or adult. To turn a cell's clock back to zero, researchers like Wolf and Schramm use a technique called nuclear transfer technology. This is the basic method that produced Dolly, Neti and Ditto, and the three identical calves. In all three cases, the scientists removed an egg's nucleus (that is, its DNA, the genetic material that makes each individual unique) and replaced it with the nucleus from another cell. For Dolly, the nuclear material came from an adult cell; that of Neti and Ditto came from two separate embryos; and the calves' was derived from the cells of a single fetus. In all cases the cuckooed eggs were then persuaded to grow and divide normally.

That's the straightforward part of the formula. In between lies a minefield of potential problems, many unique to whatever species is being cloned. "We're not following a recipe," says Wolf. The conditions under which the embryos grow vary widely: each animal has its own required temperature, for example. And an embryo's cells begin to differentiate at different moments for different species. Sheep, calves, monkeys, and humans all reach the eight-cell stage before they start differentiating, but mice begin when the embryo consists of only two cells. . . . "There's also a lot of variation among mammalian species just in the size and nature of the egg," Wolf adds. In some mammals, such as pigs, eggs are dark in color, making it hard to tell if they are viable. While that's not a problem for manipulating rhesus monkey or human eggs, where any discoloration means the egg is dead, simply getting the eggs is. "You can get buckets of eggs from slaughterhouses" for livestock species, explains Schramm. "But every egg you get from a monkey is worth its weight in gold."

Dividing an Embryo

In the case of Neti and Ditto, eggs were first harvested from several rhesus females whose ovaries had been stimulated with hormones.

"You give them hormone shots twice a day for eight days," says Schramm, and "then, if you're lucky, maybe you get 20 eggs. Out of these, 16 may be mature. And from these 16, perhaps 12 will be fertilized." The eggs are fertilized by placing them in a dish with the male monkeys' sperm, and the resulting cells are grown in a nutrient broth under what Wolf terms "well-defined conditions; this is something we know a lot about from human infertility studies and that can be applied to our monkeys." Each embryo is allowed to grow for three days, dividing into eight cells. At this stage, all the cells are still identical to one another and largely unprogrammed. "Theoretically, you could produce a complete individual from each of these cells," says Wolf, giving you eight identical monkeys. But only theoretically, because most do not survive the coming manipulations.

In the next step, the cells, called blastomeres, are carefully teased apart; they constitute the donor nuclei. "Each one," explains Wolf, "is really one-eighth of an embryo," but that one-eighth contains the key ingredient: the nuclear DNA, all that's apparently needed to get the process ticking again.

You might expect that geneticists could divide each embryo into eight blastomeres, wait for each blastomere to grow into an eight-cell embryo, and repeat the process indefinitely. But that's not possible, says Wolf, because the embryo's cells begin differentiating into limbs and organs after a certain amount of time has passed since its development began, regardless of how many cells it has. An embryo grown from a blastomere will have only an eighth as many cells to work with as an entire embryo; if you divided it again, it would have only a sixty-fourth as many cells. "As development proceeds, when time for it to differentiate arrives, it doesn't have enough cells for the job," says Wolf, and even a blastomere will be less viable than an entire embryo. Because the cues to develop come from the cell's cytoplasm—the material that fills the cell—rather than the nucleus, the blastomere's clock can be reset by transferring its genetic material to a new egg full of fresh cytoplasm.

Transferring the Nucleus

Using micropipettes, the scientists remove and discard the nuclear DNA from another batch of rhesus monkey eggs. That leaves the cytoplast—that is, the egg's membrane and the material that once surrounded its chromosomes. A donor cell, one of the blastomeres, is then placed next to the chromosome-free egg in a petri dish. "In normal fertilization, an egg is in a quiescent state at the time it is ovulated," says Wolf. "The sperm triggers the egg to be activated, and the cytoplasm starts the program of events that will lead to development. But here, we aren't giving the cytoplast any sperm, so we must artificially stimulate" the two cells with a chemical treatment. A pulse of electricity then causes the two cells to fuse, and a "reconstituted" embryo is formed. The order of these two steps, however, was reversed when the Massachusetts researchers at Advanced Cell Technology cloned their calves; and the chemical treatment was apparently bypassed altogether when Dolly was made. "It could be species differences, or it could be artifacts of the lab," says Wolf. "It's too early to say.

"Once we have the embryo, we can treat it as we do any other," he continues. "Most often we freeze them until we have a monkey ready for an implant." That's the other big hurdle—making sure that the recipient monkey is at the right point in her cycle for the embryo to take. Prospective recipients are monitored for several weeks beforehand. To do the actual transfer, a veterinarian surgically places the embryo into the monkey's oviduct. "Women have short, straight cervixes," explains Wolf, "so surgery isn't required" when embryos are transplanted at fertility clinics. "But a monkey's cervix is tortuous, and the only way we can implant the embryos is surgically, although we're trying to come up with other methods."

Turning on the Copying Machine

At the end of all this labor, only eight twins can be produced, and that's assuming that every transfer succeeds, which is "pie in the sky," says Wolf. "It's not the optimal method, although we used it to make Neti and Ditto." But Wolf wants a series of clones, and for this, he says, "we need a lot of identical nuclei." He expects to retrieve these donor nuclei from the cells of fetal monkeys, such as their embryonic stem cells (undifferentiated precursors for other cell types) or fibroblasts (the cells that form the body's connective tissue, which are commonly grown in labs). Both kinds can be propagated in large numbers in test tubes, making it possible, he says, "to produce a clone size that is infinite in number." In other words, he expects to turn out identical monkeys, like a copying machine with a jammed "on" switch. "We don't know yet if we can do this; that's what we're working on now."

And in fact, this same technique—growing a line of fetal cells for subsequent nuclear transfer—enabled researchers at Advanced Cell Technology to produce the identical calves. "It's a very efficient method for us already," says Steven Stice, the firm's chief scientific officer. "We're producing more viable embryos than we have cows to put them in." (Oddly, the company has had no luck cloning a pig. "They are very different from cattle," says Stice. "Every step has to be reevaluated. We're not sure what we're doing wrong.")

Scientists first began trying to clone animals using adult cells in 1938, when the German embryologist Hans Spemann proposed making a clone by removing an egg's nucleus and replacing it with the nucleus from another cell. Those efforts failed until the 1970s, when frogs were finally cloned via the nuclear transfer method. None of the cloned frogs, however, made it past the tadpole stage. And that's where the idea of adult cloning stayed until Dolly arrived.

"It couldn't be done; that was what everyone said," explains Wolf, "which is why this was such a revolutionary discovery." The Scottish team "found a way to reprogram that adult cell." They did so by starving the adult cells, thus inactivating them. Wilmut began with a vial of frozen cells taken from the udder of a six-year-old sheep. His team thawed them and placed them in a growth serum with only minimal nutrients for five days. "That's the trick that made all the difference," says Wolf. The adult cells were then fused with 277 different eggs. Out of all these attempts, only one lamb was born: Dolly. "That tells you that something was desperately wrong with the other 276," says Steen Willadsen, an embryologist at St. Barnabas Medical Center in Livingston, New Jersey.

Unanswered Questions

Because of this low success rate, "we're a long ways off from getting adult cloning to work on a regular basis even in domestic animals," adds Lawrence Layman, a reproductive endocrinologist at the University of Chicago. "It'd be highly unethical at this stage to try it in humans," since the probability of miscarriages and birth defects is high. Stice agrees: "It'd be complete folly. We've used hundreds of thousands of eggs in cattle over the last ten years to achieve these results. To start at ground zero now with humans would be morally wrong and misguided."

Some researchers worry too that damage from aging DNA may be passed on to the cloned infant. "It's going to be very instructive, watching Dolly age," says Julian Leakey, a biochemist at the National Center for Toxicological Research in Jefferson, Arkansas. "If she goes through puberty, she may be okay." But she might also have acquired some random genetic mutation that could lead to problems early in life. "That's the potential danger of cloning adults," says Leakey, "which is why it would be useful to do controlled tests in short-lived mammals, such as rodents, first. Then you could work out the odds of using aged tissue versus young tissue for cloning."

There are other unanswered questions. It's not clear which cells were used to make Dolly. "They don't know which cell from the udder worked, or why it worked," says Ledbetter. "That's a big gap, and it means we don't have any idea if every cell type will work or only certain ones." Some researchers even question that it was an adult cell at all: the udder cells were taken from a pregnant sheep, and fetal cells are known to circulate in a mother's body. Nor do researchers know if the serum starvation trick will work with other species.

Delayed Twins

Despite the difficulties, says Willadsen, "the technique will be—is being—perfected". . . . somewhere. And once that happens, it's only a matter of time before we see the first cloned humans—individuals who are a physical copy, or twin, of their mother or father, but separated by at least a generation. "When that first cloned child is born, not only will no one know that he or she is different," says Princeton geneticist Lee Silver, "no one will know that he or she is a clone. People will probably say things like, 'Oh, you look so much like your mother [if she was the nucleus donor],' and she'll smile. But no one will know, at least not until the kid is 16 and decides to sell her story to the tabloids for a million dollars."

From studying twins that were separated at birth and raised in different families, researchers surmise that such clones will also be likely to share intellectual abilities and personality traits with their sole biological parent. Clones may thus follow in the footsteps of their parent—but only in a very general way. "They will be separated by an entire generation," notes Sandra Scarr, a professor emerita of psychology at the University of Virginia in Charlottesville. "And as we all know, the cultural and social circumstances of the next generation are never the same as those of the preceding one. It's those social attitudes that shape a person's view of the world, including everything from how you view the stock market to the excesses of war. So the clones may be similar in intellect and personality, but their content will be different." Identical twins reared in the same house, she notes, listen to the same bedtime stories, eat at the family table together, attend the same schools, have the same friends and teachers. The clone and its parent, however, will share none of these experiences. "And these are the kinds of things that influence how one expresses one's genetic potential."

"People think it's going to be a robot or automaton," says Thomas Bouchard, who's led the long-term twin studies at the University of Minnesota. "Nothing could be further from the truth. They'll be their own persons, and that's why the idea of cloning doesn't bother me in the least. It's nonsense to be afraid of it." Yet because of this culturally ingrained idea of what a clone is, some ethicists are concerned that the parent of a clone may try to exert excessive control over the child. "Parents already control their children to an extraordinary degree," says Lori Andrews, a law professor at the Chicago-Kent College of Law. "Will these clones be held in some kind of genetic bondage to their parent? They might put undue pressure on the child to grow up in a certain way, so that it really doesn't have its own identity."

Sources of Difference
Other researchers question how similar the clones will be, even physically. "We already know from studying monkeys and children that there's considerable variation at birth," says Christopher Coe, a psychologist at the University of Wisconsin in Madison. Coe intends to explore this variation with the cloned rhesus monkeys that his colleagues Watkins and Schramm are attempting to produce. Since the

cloned embryos will be implanted in different mothers, they'll "give us the best opportunity we've ever had to clarify what we mean by nature versus nurture," says Coe. "It's the project I've dreamed about since graduate school, 20 years ago." For instance, how different will a cloned monkey that's implanted in an older mother be from one that's grown in a younger mother? "To what degree do in utero influences affect the development of the baby?" asks Coe. "And how much do the mother's actions, what she eats, and whether or not she's dominant or submissive, influence her baby's growth? The prenatal environment plays a far bigger role in shaping a baby than most people realize." The cloned monkeys, he believes, will probably look alike (although they could also differ in such things as their weight at birth) but will nevertheless "be quite different."

Other research on human twins also suggests that such things as how early the cells divide into twins and where the twins are placed in the uterus affect how "identical" they are after birth. "I think that's the real question: Just how different will these cloned babies be at birth, despite being genetically identical to their parent?" adds Coe. In an effort to establish the cloned monkeys' individuality, he will be measuring everything from their birth weight to how quickly they hold up their heads and how long they nurse.

Then too there's the question of the influence of the mitochondrial DNA. Not all of a cell's DNA is found in the nucleus; the mitochondria, tiny organs a cell uses to transform food into energy, have their own DNA. Although the donor egg will receive a new nucleus, it will retain its mitochondrial DNA, which may well be different from the donor's. "It's only a small amount of genetic variation, but it's there," says Silver, though, he adds, "there is nothing in the mitochondrial DNA that matters in making us different from each other."

In short, cloning yourself will not roll the clock back. It will not produce your soul mate and may not even give you your complete identical twin. What it will do is give you a baby that is more biologically related to you than anyone else. And that, says Silver, is why cloning will happen and few people will harshly judge those with infertility problems who choose it as a way to reproduce. "It's instinctive, I think, to want to have a biological child. That's what cloning offers—a chance for some people to have what they thought they never could have: a child of their own."

STEM CELLS: A PROMISING LINE OF CLONING-RELATED RESEARCH

Gregg Easterbrook

In December 1998, Gregg Easterbrook writes in the following article, two independent researchers announced that they each had isolated and copied a special type of cell (called a stem cell) from human embryos and grown the cells in the laboratory. Stem cells, Easterbrook explains, have the potential to develop into any kind of tissue in the body. Therefore, he relates, stem-cell duplication might be used to provide sources for replacement tissues or organs, freeing doctors and patients from dependence on a limited supply of donated organs and the need to overcome the immune system's tendency to reject organ transplants. Easterbrook points out that, promising as it is, stem-cell research raises serious ethical questions about the use of human embryos and fetuses in scientific experiments. It also may make human cloning both possible and "respectable," a thought that many view with alarm, he notes. Easterbrook concludes by considering some of the implications of cloning humans. In addition to writing frequently for *The New Republic*, Easterbrook is the author of several books, including *A Moment on the Earth: The Coming Age of Environmental Optimism.*

For John Gearhart, a biologist at Johns Hopkins University, professional life had been an exercise in slamming against walls. Gearhart's specialty is Down's syndrome, triggered when one of the infant body's chromosomes copies itself once too often. Gearhart had spent 20 years trying to puzzle out this genetic error. "All our data suggested that Down's was caused by something that happens quite early in embryogenesis," he says—but the only way to find out what happens then would be to conduct experiments on human embryos, a prospect repugnant at best. Trying to think his way out of the problem, Gearhart wondered: What if there was a way to isolate and culture embryonic "stem cells," the precursors of all body parts? If they could be transferred to the laboratory, it might become possible to study the cytology of [cell changes following] conception.

A Medical Boon?

Stem cells are the philosopher's stones of biology, magical objects capable of metamorphosing into any component of the body: heart, nerves, blood, bone, muscle. Mammal embryos begin as a clump of stem cells that gradually subdivides into the specific functional parts of the organism. Researchers have long assumed that, because stem cells are genetically programmed to change into other things, it would never be possible to control them, let alone culture them. But Gearhart and another researcher working independently, James Thomson of the University of Wisconsin, found this is not so.

In December 1998, Gearhart and Thomson announced that they had each isolated embryonic stem cells and induced them to begin copying themselves without turning into anything else. In so doing, they apparently discovered a way to make stem cells by the billions, creating a biological feedstock that might, in turn, be employed to produce brand-new, healthy human tissue. That is, they discovered how to fabricate the stuff of which humanity is made.

Researchers had already demonstrated that stem cells might be a medical boon by showing that such tissues extracted from aborted fetuses could reverse symptoms of Parkinson's disease. But so many fetuses were required to treat just one patient that the technique could never be practical, to say nothing of its harrowing character. By contrast, Gearhart and Thomson have found that stem cells can be reproduced roughly in the way that pharmaceutical manufacturers make drugs.

If researchers can convert stem cells into regular cells like blood or heart muscle and then put them back into the body, then physicians might cure Parkinson's, diabetes, leukemia, heart congestion, and many other maladies, replacing failing cells with brand-new tissue. Costly, afflictive procedures such as bone-marrow transplants might become easier and cheaper with the arrival of stem-cell-based "universal donor" tissue that does not provoke the immune-rejection response. The need for donor organs for heart or liver transplants might fade, as new body parts are cultured artificially. Ultimately, mastery of the stem cell might lead to practical, affordable ways to eliminate many genetic diseases through DNA engineering, while extending the human life span. Our near descendants might live in a world in which such killers as cystic fibrosis and sickle-cell anemia are one-in-a-million conditions, while additional decades of life are the norm.

Granted, sensational promises made for new medical technologies don't always come to pass, and some researchers are skeptical about whether stem-cell technology will pan out. But Harold Varmus, head of the National Institutes of Health (NIH), recently declared, "This research has the potential to revolutionize the practice of medicine." Notes John Fletcher, a bioethicist at the University of Virginia, "Soon every parent whose child has diabetes or any cell-failure disease is going to be riveted to this research, because it's the answer." Ron McKay, a stem-cell researcher at the National Institute of Neurological Disorders and Stroke, says, "We are now at the center of biology itself." Simply put, the control of human stem cells may open the door to the greatest medical discovery since antibiotics.

Disquieting Aspects

But there are disquieting aspects to stem-cell research, too. The first is that, for now, the only way to start the process of controlled stem-cell duplication is to extract samples from early human life. Gearhart used fetuses aborted by Baltimore women; Thomson, embryos no longer wanted by Wisconsin in vitro fertilization (IVF) clinics. Gearhart, Thomson, and other stem-cell researchers propose to continue drawing on such "resources," as some bloodless medical documents refer to the fetus and the embryo. This is possible because, even though Congress has placed a moratorium on federal funding for experimentation on most IVF embryos and most kinds of fetal tissue, no law governs what scientists can do to incipient life using private funding, either in research settings or within the burgeoning IVF industry.

Because the rules have banned embryo research by federally funded biologists, but not comparable private science, Congress has created the preposterous situation in which most stem-cell research is not being done by publicly funded scientists who must pass multiple levels of peer review and disclose practically everything about their work. Instead, most stem-cell science is in the hands of corporate-backed researchers. Gearhart's and Thomson's projects, for example, are being underwritten by Geron, a company whose name derives from "gerontology," and which anticipates a licensing El Dorado if stem-cell-based good health can be patented and sold to the seniors' market. "That a sensitive category of research is legal for people who are not publicly accountable, but illegal for those who are accountable, is just very strange," says Thomson.

But the greatest anxiety about stem-cell research is that it will make human cloning respectable. Many of the techniques being perfected for the medical application of stem cells are just a hop, skip, and a jump away from those that could apply to reproductive cloning. Society isn't even close to thinking through the legal, ethical, regulatory, and religious implications, but, thanks to stem-cell research, cloning may arrive much, much sooner than anyone expects.

Stem cells stand in the vanguard of human life. When a sperm penetrates an egg, it triggers a majestic sequence whose first step is to create a new structure that is composed mainly of stem cells. Biologists call such cells "undifferentiated," meaning they have not yet decided what they will be. Once the fertilized ovum [egg] implants in the uterus, differentiation starts. Some stem cells become placenta; others begin differentiating into the baby's organs, tissue, or blood. A stem cell might divide into any of the many components of the body, but, once it does, it can only continue growing as that part.

Because once a stem cell begins to differentiate it cannot turn back, biologists assumed that all stem cells could never turn back. But, in 1981, experimenters succeeded in extracting stem cells from the embryos of mice. By the mid-'90s, researchers had learned which chemicals instruct mouse stem cells to become particular tissue types and how to insert the new tissues back into mice. Loren Field of Indiana University became so adept at signaling mouse stem cells to become mouse heart cells that "his lab is almost pulsating with heart cells beating in dishes," Gearhart says. . . .

Regulating Embryo Research

Both Gearhart and Thomson call on Congress to enact clear legal guidelines for their field. Thomson says, "The human embryo is the most special cell in biology, and there are just some things you shouldn't do to embryos"—mainly clone them. The primary point stem-cell researchers make in their own favor is that the cells they experiment upon, once brought into the lab, might be made into muscle or blood, but can no longer become a human being. This assertion seems true, though slightly cute, since the reason the cells cease being capable of personhood is that they've been artificially snatched from it. But then no one plans to conceive the IVF embryos that Thomson gets, and the fetuses Gearhart receives have already had their lives terminated. Neither biologist can change these things, though both might change others' lives for the better.

Reflecting the delicacy of the situation, stem-cell researchers are beginning to wrestle over the terms *totipotent* and *pluripotent*. A totipotent cell is what exists at the earliest germination stage, when each stem cell is capable of becoming a whole person. A pluripotent tissue is an isolated stem cell, capable of transforming into any desired cell type, but not of becoming a whole person. Not, at least, with current technology.

In January 1999, government lawyers sided with the pluripotent versus totipotent distinction, ruling that the NIH can begin funding stem-cell research on the grounds that the cells being worked with cannot become persons and thus are not embryos. This ruling hasn't yet taken effect; assuming it does, there will be beneficial results. Publicly funded scientists from research-center universities will jump into stem-cell investigations: research-center scientists are generally the country's best, and always the most accountable. Equally important, federal funding will move stem-cell findings into the public domain rather than allowing them to become proprietary. Geron shares samples of its stem cells with academics but asks the recipients to sign a statement that Geron owns the knowledge embodied in the cell line. Once public funding flows, proprietary claims will diminish.

Bringing public funding to stem-cell research will force a public debate on this new biology. There has been little so far. In Congress, a few members, such as Representative Jay Dickey of Arkansas, have declared themselves opposed, for pro-life reasons, to any research on embryonic cells. A few members, such as Senator Tom Harkin of Iowa, have openly endorsed stem-cell studies. Senator Arlen Specter of Pennsylvania was expected in 1999 to introduce legislation making human stem-cell research explicitly legal, as it is in the United Kingdom. But Specter says he will postpone action, feeling the time isn't right. . . .

Adult Stem Cells?

There is one possible avenue of escape from the moral enigmas of stem-cell research: if stem cells can be found in adults, there will be no need to draw on embryonic "resources." Biology textbooks call this quest hopeless, since by adulthood every cell is differentiated and incapable of further transformation. At least that was the view until February 1997, when the British researcher Ian Wilmut cloned Dolly. Essentially, he did this by taking cells from an adult lamb, making them act like stem cells, and then fusing their DNA into a donor egg which germinated into a baby lamb genetically identical to the adult. Mainstream biologists had thought that embryos, still rich in stem cells, might be cloned, but never adults—or that, if adult stem cells could be found, they would be incapable of reactivating. Since Dolly, however, cows and mice have been cloned from adult cells by variations on the Wilmut technique. Researchers are now finding indications that small amounts of stem cells continue to exist, overlooked, in the adult's nerve tissue and elsewhere; it may be that there are small adult stem-cell deposits throughout the body.

One biotech company, Advanced Cell Technology, claims it has already grown human stem cells by starting with adult tissue. The company says it removed the DNA from a skin cell of a man, inserted these genes into a cow's egg cell in such a way that the human DNA took over, and then watched as the cow egg dutifully produced human stem cells. Word of this procedure, mingling human and animal reproductive cells, caused President Bill Clinton to say he was "deeply troubled" by the experiment—though, if the work was done as the company claims, the tissues created were human, since the human DNA had taken over.

Several experts dispute whether the experiment actually happened, and Advanced Cell Technology hasn't yet published the peer-reviewed data that back up a discovery claim. But whether or not stem cells have been made from an adult, some biologists are beginning to think this will happen eventually—"eventually," at the current pace of biotech advances, often meaning "next year." Stem cells derived from adults would not only resolve qualms about embryonic tissue; they might have superior therapeutic properties. Suppose you had a liver disease. If one of your own cells could be used as the template for fresh stem cells that would then be converted into liver tissue, what would end up transplanted into you would contain your own DNA and antigens, which presumably would forestall tissue rejection. Transplants might become something the typical person experienced several times in adulthood. Life expectancy would shoot upward, along with the health care share of the gross domestic product (GDP). . . .

Now back to that cow's egg experiment that may or may not have succeeded. Chromosomes from a grown person were fused into a donor egg, the egg began to germinate, and then stem cells were extracted. Suppose the stem cells hadn't been extracted. What might have developed might have been the embryo of a human clone.

A Gateway to Human Cloning

Promising as stem-cell research is, what it's doing in the larger scheme is accumulating the technical information that will make human cloning possible. In principle, stem-cell technology might allow a clinic to isolate an adult's gene endowment, engineer it for frost resistance or God knows what else, and then, by allowing the new stem cells to reproduce themselves before implantation, clone unlimited facsimiles of the person.

It may be that the knowledge of cloning is unstoppable, in the sense that no force has ever incarcerated knowledge. And cloning should not be feared in and of itself, for there are arguments in its favor. Clones would be facsimiles of their parents (technically, parent) only in the physical sense. They'd be born as babies—no imaginable technology would create life directly as adults, making the business about rich men or dictators Xeroxing themselves a Hollywood silliness that detracts from the serious arguments against cloning. Character-shaping effects of each generation's particular upbringing would inevitably make the cloned child differ from the parent, while a clone's thoughts, personality, and experiences would reflect the unique human dignity possessed by every individual. No one argues that each member of a pair of identical twins, who are genetic duplicates, does not possess unique dignity.

Human cloning would also have the beneficial effect of providing the means for even the infertile person to conceive and raise a child, a blessing it is easy for the fertile to overlook. Thus we should not be squeamish about cloning merely because it mixes reproduction, technology, and the new. Leon R. Kass has argued that cloning is horrifying in part because it would make possible "the grotesqueness of conceiving a child as an exact replacement for another who has died." Aside from the fact that child B would not and could not be "an exact replacement," as both a parent and a churchgoer this possibility does not strike me as grotesque. If a child's life were cut short, the prospect that the lost child could leave behind a similar life, consoling the parents and preserving the memory of child A, might seem like a miracle.

When IVF techniques made their debut in 1979, pundits pronounced themselves repelled by "test tube" children, and polls showed the public strongly opposed. By 1994, at least 150,000 IVF babies had been born, brightening the lives of couples who would otherwise have been barren, while public support for IVF science had become widespread. Babies made possible through this technology aren't weird or abnormal, and they are less likely to suffer from gene-defect diseases, since their embryonic chromosomes are comparatively easy to check. The first American IVF child, Elizabeth Carr, just turned 17, and she wants to be a journalist. According to *The Boston Globe*, she attends an annual reunion of IVF children and delights in holding the latest in her lap. Can anyone believe the world would be a better place if the technology that caused Elizabeth Carr had been banned?

An Interest in Being Born

Every new life has an interest in being born, which was the clincher argument for IVF, and may someday be the clincher argument for cloning. Imagine meeting a cloned person and asking, "If you had been given the choice of either not existing or being conceived as a clone, which would you have chosen?"

Of course, it's possible the cloned person would turn the question around and say, "If you had been given the choice of either coming into existence as an unpredictable one of a kind or as a clone, which would you have chosen?" Perhaps there are some people with genetic defects who would say they'd rather have been cloned from someone with problem-free DNA, but most of us would say we'd rather have our gloriously unpremeditated forms. Cloning should produce physically healthy children—but what about mentally healthy ones? It's one thing to bear resemblance to a mother or father; most children take pride in that. It may be another to be born with expectations about you preformed—to miss the chance, as Kass has beautifully said, to arrive "an unbidden surprise, a gift to the world."

If a child could only be conceived through cloning because circumstances made it impossible for the parent to have progeny any other way, then the clone's interest in being born might outweigh the psychological risks. But what about the prospective mother or father who simply wants a mirror-image child, especially if the motive is vanity? In that case the chance of harmful psychological burdens on the child might argue against allowing cloning. It is this kind of imaginable-now issue, rather than science fiction about biotech dystopias [nightmare societies created by biotechnology], that ought to govern public debate on stem cells and the prospect of cloning.

Outracing Understanding

Today's situation with stem cells and cloning might be likened to what would happen if a fleet of modern jet fighters were teleported back in time to ancient Sumeria. First, the ancients would marvel at the objects, noting their extraordinary complexity—as scientists marveled when they first glimpsed the extent of the double helix. Initially, they'd be too scared to touch, and some would argue that the gods would punish those who touched. Eventually, the fear would wane, and, by poking and prodding and pushing buttons, someone would manage to start one of the plane's engines, generating thunder and fire. At that point, the ancients would believe they had "discovered" the true purpose of the mysterious objects, and that, now being able to manipulate the planes, they had become masters of them.

Owing to the stem-cell breakthrough, there now stands the prospect that our children will not only live healthier lives but that their children will be the final generation of *Homo sapiens*, supplanted by *Homo geneticus* or whatever comes next. *Homo erectus* didn't last, so there's no reason to assume *Homo sapiens* won't ever give way to a next stage. If all goes well, the advent of control over our own cells might offer our grandchildren many things we would wish for them.

But it's all happening much, much faster than society understands. It's also happening under conditions in which we are telling ourselves that we understand genes because we have learned to make them do certain things, but we probably know little more about the totality of our DNA than would the ancient who doesn't even realize that airplanes are supposed to fly. It's time to move biotechnology to the center of the national debate, so that we can sort out its rights and wrongs before sheer technological momentum imposes an outcome upon us.

CHAPTER 2

CLONING ANIMALS

CLONING AND THE LIVESTOCK INDUSTRY

Takahashi Seiya

Takahashi Seiya, a senior researcher at the Laboratory of Reproductive Biotechnology in the Department of Animal Reproduction, Japanese National Institute of Animal Industry, describes the likely benefits that new cloning technology will bring to the livestock industry, especially in Japan. Seiya examines how Japanese researchers working in Japan and Hawaii extended the technique of cloning adult body cells—used to produce the famous sheep, Dolly—to cattle and mice. The technique may eventually allow production of better beef and dairy cattle in Japan and elsewhere, Seiya says, but first its success rate needs to be improved, and ethical questions related to cloning should also be resolved.

When Dolly was born at the Roslin Institute in Britain (reported in February 1997), she became the world's first clone produced from a somatic cell of a mature adult animal. (Somatic cells are those in the body that compose the tissues, organs, and parts other than the germ cells.) In a duplicate experiment in July 1998, a team led by Professor Tsunoda Yukio of Kinki University produced the world's first calf clones from a somatic cell of a mature cow. Soon after, many other research facilities reported successful births of calves cloned from somatic cells. Professor Yanagimachi Ryūzo of Hawaii University and his team have also reported the births of many somatic mouse clones.

Dolly's birth immediately became a major story around the world, not only because of its great scientific significance, but also because of fears that the technology might be applied to produce human clones. Nonetheless, there are great hopes that cloning will prove a key technology in furthering research and development in the medical and life sciences fields, accelerating the reproduction of improved livestock and making it possible to efficiently produce transgenic livestock.

With cloning technology, a cell is taken from an animal and from it another animal with the exact genetic makeup is produced through nuclear transfer. A male or female cow that has some valuable trait

Excerpted from "Until the Cows Come Cloned," by Takahashi Seiya, *Look Japan,* January 1999. Reprinted with permission from *Look Japan.*

can be replicated so that its use as a genetic resource is possible over a wide area and over a long time. Moreover, the uniformity of genetic qualities provided by clones is important to animal experiments, making it possible to formulate more precise experiments. Clone technology may help make possible the expression of traits hitherto buried in the genetic makeup of individual organisms, a factor important to research into both the influence of the environment on animals, into feeding regimes, and so on. Furthermore, whereas previously sperm and eggs had to be saved to preserve valuable animal varieties or lineages, new cloning technology means somatic (that is, non-reproductive) cells can also be used as a genetic resource.

The Roslin Institute and the Massachusetts University group both introduced a gene into a fetal fibroblast cell to produce a sheep or cow clone through nuclear transfer. This constitutes a new technology for producing transgenic animals.

In recent years, the concept of using cloned animals as factories to manufacture specific compounds for human medicine has attracted attention. In this bio-factory model, the gene for producing a certain human protein is introduced into a sheep or cow. The host animal then synthesizes the target protein in a bodily fluid such as its milk or blood. The human protein can be extracted and given to those humans who, due to some genetic defect, cannot produce enough of the protein themselves.

Xenotransplantation has also been in the news. Here, animals are produced through genetic manipulation to have organs that, when transplanted into human beings, are unlikely to trigger a rejection response. This is a boon for transplantation medicine, which suffers from a chronic shortage of organ donors.

Ethical Concerns

Countries around the world, however, are laying down rules for cloning because of fears over its use to clone human beings. In Japan, the Council for Science and Technology, a council under the Science and Technology Agency, announced an immediate moratorium on cloning research on 21 March 1997. Subsequently, the council addressed the matter in a report on "The Basic Plan for Research and Development in the Life Sciences."

On 28 July 1997, the council announced, "Recent technological advances in technology for the creation of biological forms through nuclear transfer technology (cloning technology) have made applications to creating human beings a possibility. Use of this technology is now being debated from a number of standpoints. Use of this technology to produce clones for livestock breeding, medical experiments, and preservation of endangered species, and use of this technology to culture human cells without producing human clones, has great significance in the areas of livestock breeding, scientific research, preservation of rare species, and the manufacture of pharmaceuticals. However, this technology must be developed in a proper manner so that, for instance, it does not directly run into ethical problems. The creation of clones of mammals must be conducted with full disclosure to the public."

Developing Technologies

Since the liberalization of beef imports to Japan in 1991, there has been greater need for strengthening the foundation of the domestic livestock industry and for cutting costs and improving meat quality. The Ministry of Agriculture, Forestry and Fisheries' (MAFF) AGROKEY 21 program promotes research and development of breed improvement and reproduction technology for cattle based on egg implantation technology. In Japan, egg implantation and related technology has contributed greatly to advances in livestock breeding and livestock improvement. Dr. Sugie Tadashi, former laboratory head in MAFF's National Institute of Animal Industry, was a world leader in developing the technology for implanting fertilized eggs into a cow without using surgical procedures (1964). In order to make livestock improvement through egg implantation even more effective, peripheral technologies—such as preservation of frozen embryos, embryo splitting, in vitro fertilization and culturing, and embryo sexing—have been developed.

The cloning of livestock using nuclear transfer technology is another of these peripheral technologies. In 1990, joint research by Professor Tsunoda (then laboratory head in the National Institute of Animal Industry) and the Chiba Livestock Experimentation Center resulted in the birth of Japan's first calf from the nuclear transfer of a blastomere from a fertilized egg (early stage of the embryo).

Research at a number of national animal husbandry experiment stations, universities, and private corporations has since then brought about many technological improvements, and 370 nuclear transfer calves have now been born in Japan. Although multiple monozygotic (derived from the same egg) calves have been produced, the success rate remains low and the greatest number of monozygotic offspring from a single egg so far is six. In order to produce more monozygotic calves, a method is being studied whereby a cell from an early-stage embryo is cultured in vitro and the resulting cells are used in the nuclear transfer procedure.

In 1997 the National Federation of Agricultural Cooperative Associations (ZEN-NOH) produced a calf at their research center using this method. The method, however, still requires perfecting. More research into the culturing of the cell from the embryo—the key to the method's success—is needed.

On the other hand, culturing of the somatic cells used in somatic cloning is possible with techniques taught to every student majoring in biology. The success of somatic cloning and the string of new achievements have signaled the downfall of the strategy of producing many clones from culturing early-stage embryo cells.

Advances in Cloning

Somatic cloning experiments began in Japan in August 1997, after the release of the report by the Council for Science and Technology. Presently, research aimed at establishing nuclear transfer technology at the fertilized egg level is underway at state-subsidized programs and at the prefectural level and will continue for the next several years.

To determine whether the production of clones from somatic cells is possible in cattle and whether the focus of research and development can be shifted from blastomeres to somatic cells, the National Institute of Animal Industry has produced a nuclear transfer embryo from a cow fetal cell and a cell from the skin of an adult bull cow and implanted the embryo into a recipient cow. The nuclear transfer embryos developed at the same rate as earlier experiments with blastomere clones. After implantation into the mother cow's womb, some of the embryos survived and developed to term. These embryo implantation experiments were conducted jointly with the Kagoshima Prefecture Cattle Breeding Improvement Research Institute. On July 24, 1998, one calf clone originating from a fetal cell was born at the institute, followed by two clones originating from skin cells from an adult cow.

Professor Tsunoda had previously overseen the birth of twin clone calves produced from oviduct cells. As of September 1998, the team had produced 10 calves from somatic cells. Between late August and September, two calves cloned from muscle tissue were born at the Oita Prefectural Institute of Livestock Industry. In addition to these, a total of 71 calf clones have been conceived at at least 20 different research institutes. However, nearly half of all clones born have died at birth or within several days. Miscarriages are also common. Clearly the risks are still too great to immediately use cloning technology for livestock production. Rather than simply reacting with either joy or sorrow to the success or failure of each experiment, scientists must move forward in their research with a clear grasp of all conditions surrounding cloning.

Progress and Problems

The many successful clonings during the one year since Dolly made history attest to the great hopes that people in the livestock industry have for cloning technology. Furthermore, it is greatly significant that since Dr. Sugie's development of the embryo implantation technology, technology transfers from research institutes such as MAFF's National Institute of Animal Industry and National Livestock Breeding Centers to animal husbandry experimental stations at the prefectural level have been steadily pursued and a nationwide system has been erected for employing technology of the highest standard. Whereas in many foreign countries cloning technology is treated as the province of private corporations, Japan may be the only example of a country where most cloning research is done by public institutions.

There are still many problems that must be resolved in order for cloning technology to be employed in the livestock-raising field. Even if the success rate were improved, there are still many issues to address before this technology can be implemented—questions such as whether clones will truly show the degree of similarity expected, and how cloning will fit into the breed improvement system. Furthermore, it is the responsibility of researchers and livestock breeding scientists to sincerely address the doubts and fears that the public still has towards cloning technology. Certainly, in resolving the problems facing this technology, it is vital that it be developed as a technology that will contribute to the good of mankind.

"PHARMING" CLONED ANIMALS

Elizabeth Pennisi

In the following selection, Elizabeth Pennisi describes plans to use cloning technology to produce herds of genetically identical livestock. Pennisi reports that it is already possible to transplant selected human genes into animals such as sheep and cattle, resulting in transgenic animals able to produce human blood factors or other pharmaceutical substances in their milk. However, she points out, this technique is difficult, and cloning could feasibly increase its yield by allowing a single transgenic animal to be duplicated repeatedly. On the other hand, Pennisi remarks, the success rate of animal cloning is still very low; unless the success rate rises, intensive cloning of transgenic livestock may not be practical. Pennisi is a frequent contributor to *Science* magazine.

Births are usually announced on a newspaper's society or personal pages, not on the front page. But that convention didn't apply to Dolly and Polly in 1997 and George and Charlie in January 1998. These white-faced lambs and Holstein calves made headlines as the products of cloning technologies that have generated fascination and fear—a reaction fanned in January 1998 by the improbable claims of a physicist [Richard Seed] who says he plans to clone adult humans within 2 years. But the technologies have done more than spawn an ethical debate about the prospects for human cloning: They have also galvanized efforts to create transgenic livestock that will act as living factories, producing pharmaceutical products in their milk for treating human diseases and, perhaps, organs for transplantation.

That was always the main intention of Dolly's creators, Ian Wilmut, Keith Campbell, and their colleagues at the Roslin Institute and PPL Therapeutics in Roslin, Scotland. But in the year since the announcement of Dolly's birth, a dozen other groups have been adapting the technique used by the team in Scotland. Some want to clone animals bearing working copies of transplanted genes. Although key problems remain to be solved, these efforts—many of which were reported in January 1998 at the annual meeting of the International Embryo Transfer Society in Boston—have already resulted in the birth of sheep containing a human clotting factor gene and calves containing foreign marker genes. Experiments in which the nuclei of pig cells have been fused with cow eggs have also given tantalizing results.

Brave New "Pharm"

This work is invigorating the "pharming" industry: Underwriting the cloning frenzy are biotech and pharmaceutical companies eager to cash in on its potential for creating transgenic livestock. "There is a huge industry that is organizing itself around [the new cloning] technology," says James Robl, a developmental biologist of the University of Massachusetts, Amherst.

There is, however, a crucial difference between these experiments and the original Dolly breakthrough—a distinction that has sometimes been lost in the public discussion of the implications of these new results. Dolly was cloned by taking nuclei from adult mammary gland cells, starving them of nutrients to reset their cell cycles, then fusing them with sheep eggs whose own nuclei had been removed. But this procedure was very inefficient—producing only one success out of the 277 eggs that took up the new DNA. The later experiments all use nuclei from fetal cells, which have proved more efficient at generating viable offspring than adult cells. Indeed, so far the Dolly experiment has not been exactly replicated, and some scientists have even questioned whether Dolly is in fact the clone of an adult.

Animal geneticists have jumped on the technology because it potentially offers a far more efficient way to produce transgenic animals than previous techniques, which involve the injection of foreign DNA into newly fertilized eggs. The success of an egg injection is not known until after the offspring is born. For example, using egg injection, PPL Therapeutics took years to develop a flock of 600 transgenic sheep, as only about 4% of the lambs carried the desired gene.

In contrast, nuclear transfer technology allows researchers to select as nucleus donors only those cells that express the transplanted gene. Moreover, in theory, those cells could provide as many clones as needed in a single generation. "In one fell swoop, you get what you want," says PPL research director Alan Colman. Indeed, Will Eyestone of PPL's Blacksburg, Virginia, facility told the society's meeting that egg injection "may well become old-fashioned."

From Sheep to Cows to Pigs

Campbell, who recently moved from the Roslin Institute to PPL's labs 300 meters down the road, Wilmut, and their colleagues were the first to announce that they had been able to produce transgenic animals with cloning technology. They reported in December 1997 that they had produced three cloned sheep, two of which are still alive, carrying the human factor IX clotting protein.

Now, Advanced Cell Technology has achieved in cows what the

team in Scotland did with sheep: Robl and his colleague Steven Stice announced at the International Embryo Transfer Society meeting the birth of two calves carrying a foreign gene. To produce these transgenic animals, the researchers first grew bovine fetal fibroblast cells in the laboratory and then added an antibiotic-resistance "marker" gene. Only the cells that took up the gene survived exposure to an antibiotic added to the culture dishes. The researchers then fused nuclei from the survivors with enucleated cow eggs, employing a variation on the technique used by Wilmut's group. About 40% of the resulting embryos continued to develop once inside foster mothers, and two calves—George and Charlie—were born in mid-January 1998. A third was born shortly afterward, and more are on the way. "[They are] the first transgenic cloned calves, and that's great," says Campbell of PPL, which is also doing nuclear transfer work in cattle. The three calves show "the phenomenon and the technology are not restricted to one species," adds nuclear transfer pioneer Kenneth Bondioli of Alexion Inc., a biotech company in New Haven, Connecticut.

That demonstration has been eagerly awaited. Transgenic cows, which produce 9000 liters of milk per year, should be better factories for therapeutic proteins than sheep or goats. "Milk is cheap, and we have an incredible dairy infrastructure," points out Carol Ziomek, an embryologist with Genzyme Transgenics in Framingham, Massachusetts.

Indeed, that potential has already spurred a gold rush. In October 1997, Genzyme Transgenics awarded Advanced Cell Technology a 5-year, $10 million contract to develop transgenic cows that will produce albumin, a human blood protein used in fluids for treating people who have suffered large blood losses. And in January 1998, Pharming Holding N.V. in Leiden, the Netherlands, formed an alliance with ABS Global, an animal breeding company in DeForest, Wisconsin, and its spin-off company, Infigen Inc., to develop transgenic cattle that produce the human blood proteins fibrinogen, factor IX, and factor VIII in their milk.

Other efforts are aimed at expanding the utility of pigs, particularly in biomedicine. A few companies and research groups hope to use pig organs or tissue to help meet the large unfilled demand for transplant organs. The goal is to genetically modify the animals' tissues so they are less readily rejected. Also, because a pig's physiology is more like a human's than is a mouse's, some animal scientists argue that pigs could be good models for studying human diseases if their genetic makeup could be modified so that they develop appropriate symptoms.

But this work has been lagging, partly because researchers have had trouble getting pig oocytes to start dividing after the nuclear transfers. Moreover, researchers are still working out a suitable way to keep new embryos alive until they can be placed into female pigs for continued development.

At the January 1998 meeting, several teams reported progress solving these problems. At the University of Missouri, Columbia, Randall Prather has worked out a new way to activate cell division using a chemical called thimerosal as the initial trigger. And reproductive physiologist Neal First's group at the University of Wisconsin, Madison, offered a more radical potential solution: Avoid the hard-to-activate pig egg altogether by transferring nuclei from adult pigs into bovine oocytes. "Instead of using a pig oocyte, perhaps, you could use a sheep or cow oocyte," Robl suggests. It is unclear, however, whether such cross-species embryos would ever come to term.

Reducing the Body Count

In spite of the rapid advances in nuclear transfer since Dolly's debut, some big obstacles still remain. At each step along the way some— often many—individuals don't survive. That low efficiency doomed an earlier version of nuclear transfer when it made its commercial debut a decade ago. At that time, several companies, including Granada Inc., based in Houston, were going great guns using nuclei from very early embryos to clone hundreds of calves to make large herds of genetically superior beef cattle. But by 1991, Granada had shut its doors. "We couldn't make as many calves as we wanted to," recalls Bondioli, who worked there. And too often, calves were oversized and unhealthy, with lungs that were not fully developed at birth.

Researchers see the same trends in the few cows and sheep produced by the newer cloning procedures. Large numbers of deaths occur around the time of birth. For example, PPL and Roslin lost eight of 11 lambs in their first experiment with transgenic clones. But it's not the nuclear transfer procedure itself that's at fault, says Robl. Animals produced by in vitro fertilization and other procedures involving the manipulation of embryos have similar problems, albeit at a lower frequency.

"Something that you do to the embryo . . . leads to a problem 9 months later," says George Seidel Jr., a physiologist at Colorado State University in Fort Collins. His data and other observations suggest that in problem calves the placenta does not function as it should. As a result, cloned calves have too little oxygen and low concentrations of certain growth factors in their blood.

While some researchers are experimenting with different nutrient solutions or making other subtle changes in their nuclear transfer techniques to make embryos and newborns thrive, others are frantically trying to hone the genetic manipulation techniques. Researchers currently have no control over where the foreign genes end up in the chromosomes or how many copies of the gene become part of that cell's genetic repertoire.

Developing that control would enable them to knock out specific genes, say the one encoding the pig protein that elicits a strong, immediate rejection response to pig organ transplants. "The Holy Grail for many is finding a way of getting targeted disruption of genes in livestock as we have in mice," explains Colman, who is confident that even this tough molecular biology problem will be solved quickly. "I expect we'll have targeting solved by next year," he predicts.

Such confidence is required in this fast-moving field, in which progress generally comes through trial and error. Understanding how it all works, say these scientists, will come later. "[There] clearly is at this point in time a pushing forward of the technology," says Alexion's Bondioli. "Have we learned any more biology? Probably not. But [we] have opened up a means to study [it]."

A Demand for Proof

In a perfect world, important scientific discoveries are impeccably documented and quickly replicated. But two prominent biologists say that was not the case for Dolly, arguably the most famous lamb in history because she was reportedly cloned from adult cells. Vittorio Sgaramella from the University of Calabria in Cosenza, Italy, and Norton Zinder of Rockefeller University in New York City ask for more convincing evidence that the experiment that produced Dolly worked as claimed. If in fact it hasn't, it would mean that human cloning, which for most conceivable purposes would start with adult cells, is not the immediate threat some worry about.

Because the mammary cells used to produce Dolly came from a pregnant ewe, Zinder and Sgaramella question whether she might have been cloned not from an adult mammary cell but from a contaminating fetal cell. And while Ian Wilmut, the embryologist at the Roslin Institute in Roslin, Scotland, where Dolly was cloned, and his colleagues cite evidence that that could not have happened, they may never be able to prove their assertion conclusively. Because "none of us expected to get Dolly," says embryologist Alan Colman of PPL Therapeutics in Roslin, which collaborated in the work, "we didn't do what we should have done" to document the genetic composition of either the ram that impregnated the ewe or the fetus she carried. Consequently, Dolly's DNA can't be compared with theirs. But the Roslin group also says that some of the other data Zinder and Sgaramella want, concerning whether Dolly's DNA has the mutations and other changes expected in an adult, will be available as soon as the analyses are completed.

Still, even if Dolly is an adult clone, no one has yet exactly replicated the experiment that produced her. Few laboratories, Wilmut's included, have even tried, mainly because the emphasis now is on using DNA from fetal cells, rather than adult cells, to commercialize the nuclear transfer technology used to create Dolly. Others, including Neal First's team at the University of Wisconsin, Madison, and Mark Westhusian's group at Texas A&M University in College Station, have tried to clone cows from adult cells but failed. None of the embryos survived to birth, they note. Similarly, in Boston at the annual meeting of the International Embryo Transfer Society, a team of researchers from Germany and Austria reported it had tried to use heifer udder cells as nuclei donors, but no embryo lived past 40 days of gestation.

Two other teams, one led by James Robl and Steven Stice, developmental biologists at the University of Massachusetts, Amherst, and the other at the biotech firm Infigen in DeForest, Wisconsin, say they have calves in utero that were cloned from adult cells. However, neither team is confident enough that these calves will make it through the final months of their 9-month gestation to reveal the tentative due dates.

But other results from the First team support the Roslin group's finding that adult DNA can be induced to support embryonic development. They used cow oocytes as universal recipients for nuclei obtained from the ear cells of adults from four other species: rats, sheep, pigs, and monkeys. Although only about 34% of eggs receiving rat nuclei began dividing, almost 86% of those with monkey DNA and 52% with pig DNA were activated, Wisconsin's Maissam Mitalipova reported at the embryo transfer meeting.

Dividing eggs continued to develop, with many expanding to 130 cells and reaching the stage where they needed to be implanted in a womb. These developing embryos also contained a protein, not found in the ear cells themselves, that is usually produced only in cells capable of developing into whole new organisms. "It means that if you can do [nuclear transfer] with fetal fibroblasts, you can do it with adult cells," says First. Robl agrees. Getting another clone from an adult cell is "a matter of time," he says. "If you do enough [transfers] and get lucky, you can do it."

CLONED ANIMALS MAY SUFFER

Meg Gordon

Most descriptions of recent advances in the cloning of animals describe the technology's benefits to medicine or its potential profitability but do not consider its possible harm to the animals themselves, Meg Gordon claims. Few laws protect cloned or transgenic animals from abuse or health risks, she says, yet those procedures can be very harmful to animals. For example, Gordon notes, a type of pig genetically altered to produce leaner meat suffered from severe bone and joint problems. She quotes several scientists who believe that more research should be done to see whether cloned animals will suffer as a result of gene alteration. Gordon proposes that animal welfare regulations in countries such as Britain and the United States be tightened before research on genetic alteration and cloning of livestock proceeds. Gordon is a correspondent for *New Scientist*, a British science periodical.

The flock of sheep grazing in a field on a farm outside Edinburgh in Scotland looks and behaves just like any other flock of sheep. But these animals are highly unusual. They belong to PPL Therapeutics, a fledgling biotechnology company, and they have been genetically engineered to secrete in their milk a protein called alpha-1-antitrypsin, which helps to treat cystic fibrosis.

The idea of animals being manipulated to produce substances useful to humans conjures up images of hi-tech efficiency. Yet in reality it is a pretty hit-and-miss affair. PPLs flock of sheep contains both high and low yielders of the precious product, and for each transgenic sheep created there are numerous expensive failures. Unfortunately for PPL's shareholders, these limitations mean that the flock can only ever supply a fraction of the £250-million world market for alpha-1-antitrypsin.

High Hopes for Cloning

But if future generations of these walking pharmaceuticals factories were cloned from the current herd's top producer using the technique

Excerpted from "Suffering of the Lambs," by Meg Gordon, *New Scientist*, April 26, 1997. Reprinted by permission of *New Scientist*.

that created Dolly the sheep in February 1997, the company's share of the market would skyrocket.

In the furore which followed the announcement that Ian Wilmut, Keith Campbell and others at the Roslin Institute and PPL Therapeutics had cloned a sheep from an adult cell, the scientists tried to defuse public concerns that their technique could be used to clone humans by listing the benefits that cloned livestock could offer to medicine. They talked of a future in which herds of identical cows supplied lavish amounts of medically important proteins, of sheep with cystic fibrosis and other diseases which would help doctors find new cures, and of cloned transgenic pigs that could help to meet the desperate shortfall in human organs for transplant.

However, now that the dust has settled, many are beginning to ask whether such a future would be as rosy as Wilmut and his colleagues would have us believe. Now that cloning has the potential to turn a rare experimental procedure—the creation of transgenic animals— into a profitable, industrial process, ethicists, geneticists, agriculturists and animal welfare activists are warning that the new technology could encourage serious abuses of animal welfare. They point out that although the first transgenic farm animal was created in 1985, issues such as how to minimise suffering and how to police the production of engineered animals remain unresolved. It would be dangerous, they say, to allow the cloning of transgenic animals without first tightening up animal welfare regulations.

Legal and Ethical Free Fall

"Where transgenics and clones are concerned, it is legal and ethical free fall," says Andrew Kimbrell, a lawyer with the International Center for Technical Assessment in Washington DC, which monitors the use of new technologies. Bob Combes, a geneticist and toxicologist at the University of Nottingham Medical School in Britain, who also works with the Fund for the Replacement of Animals in Medical Experiments (FRAME), is calling for an international committee to be set up to look at the welfare issues surrounding transgenic animals. He would like to see regulations which prevent companies from developing herds of transgenic animals until the long-term effects of each foreign gene on the animals' health have been fully assessed.

"There are insufficient controls," says Combes. "People argue that these animals are so fantastically important and the benefits so profound that they should not have to go through the cost-benefit analysis and weigh animal suffering and other ethical concerns in the equation. It's the technology taking over, and this is wrong." Caren Broadhead, scientific officer for FRAME, says that genetic engineers "have no idea how [transgenic] sheep could suffer".

Current laws on transgenic animals are remarkably nonspecific. In the US, once an animal has been engineered to produce a protein that is to be tested as a medicine, its welfare is largely regulated by the Food and Drug Administration (FDA) under the same laws that would govern a vat of cells. "If an animal is used as a 'bioreactor', the animal is the source of manufacture, and the FDA would regulate," says biotechnologist Frank Tang of the Department of Agriculture Animal and Plant Health Inspection Service in Riverdale, Maryland.

In addition, there are no safeguards in the US to prevent a company from creating large numbers of transgenic animals before it is certain that the foreign gene will not harm the animal or its offspring. The regulations in the European Union are just as vague.

Cloning Transgenic Animals

Using transgenic animals to manufacture useful proteins still remains inefficient. Out of 10,000 eggs injected with foreign DNA, only about three make it to adulthood and produce the desired protein in sufficiently high quantities. The techniques used to create Dolly offer two potential shortcuts. The pharmaceuticals companies could create just one good transgenic animal by conventional techniques and then clone it ad infinitum to create flocks with a human disease such as cystic fibrosis for drug testing. Or, because Dolly's genetic material came from cultured cells from adult sheep, the genetic manipulation could be done in these cells. This could allow geneticists to be more precise about the changes they are making, enabling them to introduce and remove genes at will.

Wilmut and his colleagues readily acknowledge that they have a few more hurdles to clear before the two technologies—cloning and transgenics—can be combined. The technique that created Dolly must be repeated and made more efficient, they say. Campbell points out that just one out of 277 egg cells successfully took up the adult DNA. And no one has yet used the adult cloning technique in a species other than sheep.

But most agree that the financial rewards will be sufficient incentive to overcome these barriers. "The whole reason [for] cloning is to make it a whole lot easier to create transgenic [animals] that produce valuable pharmaceuticals," says physiologist and cattle rancher George Seidel of Colorado State University in Fort Collins, who studies the possibilities for cloning livestock. And there are plenty of successful transgenic animals that would make suitable candidates for cloning, say scientists. According to Carl Gordon, a biotechnology analyst for financial consultants Mehta and Isaly in New York, genetic engineers have created no fewer than 45 transgenic goats, cows, pigs and other livestock that secrete everything from human antithrombin III, a protein that helps to stop blood from clotting, to human prolactin, which boosts the immune system.

Although most welfare concerns are over the creation of transgenic animals rather than the cloning of those animals, the new technology has itself thrown up important issues. For instance, cloning by embryo division has a tendency to create sheep and cows that are born up to twice the normal size. This strange phenomenon has already led to the downfall of one cow cloning company, Granada Genetics of Houston, Texas, because the mother cows could not deliver their calves.

Transgenic mistakes can be unpleasant for the animal. One example is the infamous "Beltsville pig", which was engineered by researchers at the US Department of Agriculture in Beltsville, Maryland, to produce human growth hormone in an effort to stimulate growth and reduce fat on the animal. The hormone succeeded in making the pig grow faster without extra food, but it suffered terribly from side effects including severe bone and joint problems.

Protected by Investment

Some scientists dismiss concerns over the threat to animal welfare posed by cloning. Robert Foote, professor emeritus of animal science at Cornell University in Ithaca, New York, insists that cloning transgenic farm animals would be a good thing. Producing medicinal products in milk is "an excellent use" of animals, he says. He argues that the large amounts of money invested in the development of transgenic animals gives them a measure of protection, as do laws designed to ensure the humane treatment of farm animals and animals used in experiments, such as the US Animal Welfare Act and the British Animal Scientific Procedures Act.

But others insist that these laws are not comprehensive enough to prevent the welfare of animals produced by genetic engineering and cloning from being abused. Charles McCarthy, senior research fellow at the Kennedy Institute, Georgetown University, suggests that the only way to ensure abuses are avoided may be to have transgenic animals monitored constantly "by someone knowledgeable about the species who will recognise signs of neurological disorders and behavioural changes that may indicate suffering".

A Need for Genetic Diversity

Keay Davidson

Keay Davidson, chief science writer for the *San Francisco Examiner*, points out that although cloning of farm animals offers potential benefits, it presents dangers as well. According to Davidson, the livestock industry hopes to use cloning to produce an "ideal" animal from some standpoint (for example, a cow that yields a high percentage of lean meat). Once such an animal is produced, Davidson says, the profit-oriented livestock industry will be tempted to clone endless genetically identical copies of it. The result could be a significant loss of genetic diversity, the author warns, which would put herds of cloned animals at risk of being wiped out by a single disease that they are not genetically resistant to. Lack of genetic diversity resulted in great crop losses during the Irish potato famine in the early nineteenth century and an epidemic of corn blight in the United States in 1970, Davidson writes. In order to avoid similar catastrophes, the author concludes, both the livestock industry and the government need to make a commitment to preserving genetic diversity.

Unless carefully monitored, cloning—the Xerox-copying of life, as it were—could accelerate one of the scariest trends in U.S. agriculture: the shrinking variety of many livestock and crops.

Shrinking variety? Those words may sound strange to anyone who strolls, eyes wide and mouth watering, through a supermarket. The display cases are commercial cornucopias, packed with foods from around the globe.

But the rich display camouflages creeping homogeneity in genetics. Many U.S. crops and livestock come from an ever-shrinking number of genetic types. Big agribusinesses like limited types because they're easier to mass-produce—like Model Ts or microchips.

And if mass-cloning proves feasible and affordable, perhaps decades from now, then the shrinking genetic variety of U.S. farms can proceed to its logical conclusion—to what one might call the "Brave New Farm." There, all crops and livestock would be "identical twins," where every chicken or cucumber has exactly the same genes as every other chicken or cucumber.

Excerpted from "Vulnerable Farm Species," Keay Davidson, *Washington Times*, March 9, 1997, as reprinted from the *San Francisco Examiner*. Reprinted with permission from the *San Francisco Examiner*.

The Risks of Sameness

Genetically homogeneous species are especially vulnerable to sudden environmental changes: say, a mutated virus or climate shift. One of history's worst disasters was the Irish potato famine of the early 19th century. It struck after the impoverished Irish grew dependent on a single variety of potato. When a potato disease wiped out the crop, hundreds of thousands of people died. Countless more fled to the New World.

In the 1970s, a USDA pamphlet recalls, "a disease called Southern corn blight swept through cornfield after cornfield from the Southeastern United States into the Great Plains. This episode cost farmers 15 percent of the corn crop that year. The 1970 epidemic reminded us of how vulnerable modern agriculture has become."

If farmers continued growing a diversity of genetic types, they would have alternate breeds to fall back on during a potato famine–type crisis. But diversity has waned over the past century in the United States: 91 percent of the different breeds of corn have disappeared, along with 95 percent of the varieties of cabbage, 94 percent of peas, 86 percent of apples and 81 percent of tomatoes, according to plant pathologist Jane Rissler of the Union of Concerned Scientists in Washington.

Many livestock also have suffered a loss of diversity. For example, most modern U.S. dairy cows belong to a single breed, the familiar black-and-white Holsteins. In the 1930s, "there were 30 breeds of pigs listed as commercial breeds," said Don Bixby, director of the 4,000-member American Livestock Breeds Conservancy. "Now we're talking about maybe eight or 12 breeds total. And only three make up more than 75 percent of all the registered purebred pigs in the United States."

Potential for Misuse

By itself, cloning has exciting potential. It could be used to mass-produce drugs, for example. It might even help preserve biodiversity by cloning species of animals that otherwise would become extinct, says Frances Katz, editor of *Food Technology Magazine*, published by the Institute of Food Technologists in Chicago.

But the potential for misuse is real. Agribusinesses face intense economic pressure to develop the "ideal" pig or cow or head of corn. For example, the ideal head of corn would taste wonderful, be the "right" color (whatever appeals to the most consumers), and be just the right height to feed into harvesting machines. Likewise, the ideal pig would yield delicious bacon, never get sick, have an amiable temperament. Once the ideal crop or livestock is found, the pressure to mass-produce it is irresistible, even if this drives out other breeds. Agribusiness, Bixby said, "is becoming more and more vertically integrated, and the people making the decisions are the accountants—the bean counters."

Older, less "ideal" varieties go extinct, as surely as the dodo, and the available genetic pool shrinks. Agriculture risks losing all kinds of genetic traits that no one cares about today, but that could prove invaluable years from now.

"If you pick an animal that you think is 'superior' today, you may be giving up on something that is better tomorrow," notes Caird Rexroad, research leader at the gene evaluation and mapping lab at the USDA's Agricultural Research Service in Beltsville, Md.

Preserving Diversity

The government has tried to preserve less popular plant seeds in "seed banks." But the banks are chronically underfunded and sometimes mismanaged, Paul Raeburn reported in his 1995 expose, *The Last Harvest*. In 1990, Congress moved toward establishing genetic "banks" of animal embryos, but the program has gone nowhere because of budget cuts.

For now, because of its great cost and complexity, agriculture experts don't see cloning as a near-term threat to the genetic integrity of U.S. agriculture. Trouble could arise, though, if agriculturists find ways to mass-clone livestock and crops as cheaply as Silicon Valley makes microchips.

The threat is not cloning itself but "economic pressures for uniformity of product," Caird Rexroad said. "We probably should be very cautious in tying up too much of our gene pool in clones."

What's the solution? The White House should at least declare a "moral commitment" to genetic diversity in agriculture. Working with Congress, it could come up with financial incentives to encourage farmers to enrich their genetic harvest—say, by raising more than one type of pig or corn.

Vice President Al Gore, who warned about "genetic erosion" in his book *Earth in the Balance,* should hold public meetings with agribusiness officials to discuss cloning and how to use it responsibly. He should ask them, in front of reporters and TV cameras, to promise never to put profits ahead of the richness of DNA on our farms. The White House has already backed genetic diversity in foreign rain forests; why not in U.S. croplands and barnyards?

CAN CLONING SAVE ENDANGERED SPECIES?

Jon Cohen

Jon Cohen, a frequent contributor to *Science* magazine, writes in the following selection that the technique of cloning from adult body cells offers great hope to zoos trying to preserve endangered species. Some zoos already have banks of frozen tissue from rare species that could provide cells for cloning, Cohen explains. Many technical hurdles need to be overcome before the technique can be considered reliable, however, he notes. Furthermore, he relates, some biologists fear that a focus on cloning, which is sure to be expensive, may divert needed funds away from more important conservation efforts such as habitat preservation. Cloning, Cohen writes, may be most useful for species whose remaining population numbers are very small or that seldom or never breed in captivity.

When Kurt Benirschke launched a program at the San Diego Zoo in 1975 to freeze cells from endangered species, he assumed that his colleagues would use the collection to unravel complex issues such as the genetic similarities among animals. Never did he imagine that scientists might one day pluck cells from the "frozen zoo" to grow new animals from scratch. But since February 1997, when researchers in Scotland reported they had cloned a lamb named Dolly from the cells of an adult sheep, the notion of cloning a Przewalski's horse, Sumatran rhinoceros, or one of the other rare species whose cells are banked at the San Diego Zoo's Center for Reproduction of Endangered Species (CRES) has suddenly left the realm of science fiction.

"The possibilities for zoos are enormous," says Benirschke, a reproductive biologist who now is vice president of the zoo. Like other zoologists, he recognizes that many scientific hurdles stand between a fibroblast—a tissue-repairing cell that makes up the bulk of the frozen zoo's collection—and, say, a healthy infant rhino. But he thinks the field has seen so many remarkable advances in recent years that the obstacles, for some species at least, are likely to fall. Says CRES geneticist Oliver Ryder, "I think [cloning] is going to produce a paradigm shift. It offers the potential for a better safety net than we thought we had." Adds Benirschke, who began working with colleagues in China after Dolly's creation to save cells from the endangered Yangtze River dolphin, "I would love to excite the international community to save as many cells as they can from as many animals as possible."

But even if the technical hurdles do fall, many conservation biologists argue that efforts to clone endangered species would be so expensive that they could derail other conservation efforts. "In the end, the very finite resources that conservation has are better directed elsewhere," contends Michael Bruford, a molecular geneticist at the Zoological Society of London's Institute of Zoology. Adds David Wildt, head of reproductive physiology at the U.S. National Zoo's Conservation and Research Center in Front Royal, Virginia, cloning should be viewed only as a "last, desperate attempt to try to preserve a given species."

Benefits of Cloning

Ryder argues, however, that cloning may offer benefits that are not immediately obvious. When people think of cloning, they often imagine legions of genetically identical individuals. But Ryder contends that the technology actually could be used to increase the genetic diversity of a dwindling species—a proposition that has taken some of his colleagues by surprise. Population geneticist Robert Lacy of the Brookfield Zoo in Illinois, for instance, says he was skeptical that cloning could enhance genetic variability, which, he notes, is "the primary thing we're trying to do with endangered species." But he was persuaded, he says, after reading Ryder's ideas on a private Internet chat group for population biologists.

Ryder reasons that for species that are down to just a few surviving individuals, clones grown from frozen fibroblasts could provide an invaluable source of "lost" genes. Suppose scientists could clone Asian wild horses, South China tigers, or Spanish ibex from cells in the CRES collection that were gathered from long-deceased animals, says Ryder. The clones theoretically would then be able to breed, reintroducing the lost genes back into the population. "It might allow you to go back and recover the genetic diversity," he says.

Ryder also argues that cloning could be an especially useful tool for biologists trying to save species that don't breed well in captivity, such as giant pandas. The more offspring an animal has, says Ryder, the more of its genome it will pass on. If a giant panda in a zoo has only one offspring, one half of the panda's genes are lost. But if biologists could clone the panda 10 times and each one produced an offspring, in effect, the original panda would have produced 10 offspring, and fully 95% of its genetic information would have been "captured." (The equation is $1 - 1/2n$, where n equals the number of offspring.)

Cloning might even serve a useful purpose with species that have never bred in captivity, such as the giant armadillo, by allowing biologists to asexually reproduce the creatures. This scenario, which would require implanting a cloned embryo of a giant armadillo in a more common relative, adds to the already formidable list of scientific obstacles. Still, says Ryder, "it could possibly guarantee genetic immortality."

When Cloning Should Be Used

In a commentary in *Zoo Biology*, Benirschke and Ryder contend that if cloning endangered species does become a reality, zoos may one day be able to breed fewer animals and retain smaller herds without losing genetic diversity. This is an important advantage, they argue, because most zoos already are short on space.

Intrigued as he is by these ideas, Brookfield's Lacy says cloning is so expensive and technically challenging that it should be used only with "a fairly narrow window" of species, those with "five, 10, or 15 animals." In most cases, he says, "with a little foresight, we'd be able to set up a breeding program that didn't cost millions."

The National Zoo's Wildt concurs, adding that lower tech, "assisted breeding" methods such as artificial insemination can often achieve the same goals as cloning. A few years ago, the black-footed ferret, for example, was down to as few as six individuals. But in 1996, Wildt and his colleagues successfully used artificial insemination to birth 16 kittens. He stresses, though, that even something as well understood as artificial insemination can be a big challenge in a new species. "We do a lot of work with assisted breeding," says Wildt. "What we've learned from working in this field for 20 years is it's really difficult."

Michael Soule, an emeritus population geneticist at the University of California, Santa Cruz, worries that cloning endangered species could distract people from saving habitats. "I don't want people to think that [cloning is] a solution to a major problem," says Soule. He heads the Arizona-based Wildlands Project, which aims to improve habitats in North America. "We've only got a few years before most of the biodiversity on the planet goes down the sink."

As they explain in their *Zoo Biology* commentary, Ryder and Benirschke do not want cloning "to minimize or supplant" current conservation efforts. "This discussion is not being advocated in lieu of saving species the only way they can be saved—in their habitats," says Ryder. . . .

Ryder and Benirschke urge their colleagues to think seriously about cloning's potential. "The future for clonable species would clearly be better than that for animals that cannot be cloned," they conclude in their *Zoo Biology* commentary. Surely, that's a definition of "fit" that Charles Darwin never imagined.

Would-Be Cloners Face Daunting Hurdles

It took Ian Wilmut and his colleagues at Scotland's Roslin Institute 277 attempts to clone one lamb, the now-notorious Dolly, from adult mammary cells. For conservation biologists who ponder the possibility of applying this advance to endangered species, those 276 failures in sheep—a species whose reproductive biology is well understood— only underscore the technical hurdles they face. Even Wilmut, who first published his results in the 27 February 1997 issue of *Nature*, points out, "The success rate is so low that you would do better to breed naturally. You would get far more offspring!"

The first challenge, says Oliver Ryder, is to see whether fibroblasts— cells made during wound healing—could be used instead of mammary cells. This question is critical because CRES's collection of cells—the world's largest—is made up of fibroblasts, stored in liquid nitrogen.

Assuming fibroblasts from adult animals could work, researchers face another challenge: harvesting eggs in a "ripened" state during ovulation. The Scottish group made Dolly by planting mammary cells from one sheep into another animal's egg that they had modified by scooping out its gene-carrying nucleus. Harvesting ripened eggs from sheep is routine because the animal's reproductive cycle is well understood. But plucking eggs from, say, a Sumatran white rhino is quite another matter, says David Wildt. "We know basically nothing about their reproductive physiology," says Wildt. "You'd have to have a rhino docile enough to allow ultrasound [to know when it is ovulating]." And once eggs are harvested, Wildt notes, different species usually require different nutritive media in laboratory cultures—media that scientists have yet to define for most endangered species.

Now, assume the transfer of fibroblasts into enucleated eggs worked and embryos developed. The next challenge would be implanting an embryo into a female that could carry it to term. Reproductive biologists say they would prefer to use females of a related, unendangered species as surrogate mothers so that females from the highly endangered population would be available for natural breeding. But it is not at all clear that the placenta carrying genes from the fibroblasts of a Rwandan mountain gorilla, for instance, would take in the uterus of a captive gorilla of a different sub species. "I think it likely that there are sufficiently specific factors to limit mixing," says Wilmut.

Kurt Benirschke, who started the CRES collection, notes that some such transfers have worked. For instance, Douglas Antczak, a veterinarian at Cornell University in Ithaca, New York, and W.R. Allen at the Thoroughbred Breeders Association in Suffolk, England, have successfully grown a zebra embryo in a horse. Antczak suggests that others might build on these results by implanting into a horse an embryo from the endangered Przewalski's horse. "It would be a good example species," says Antczak.

Betsy Dresser, a reproductive physiologist at the Audubon Center for Research of Endangered Species in New Orleans, suggests that

many of the next steps may be taken by researchers who work with
domestic animals. "The domestic-animal field has tons of animals to
work with, and money," she says. And Dresser says if they make head-
way, conservation biologists will surely take advantage of cloning.
Says Dresser, "If we can use it as a tool to save an endangered species,
you'd better believe we will."

CLONING AN EXTINCT ANIMAL

Richard Stone

Richard Stone, a deputy news editor at *Science* magazine, made a difficult journey to Siberia to research the following article for *Discover*. The Japanese scientists whose work he describes in the article have made equally challenging trips there in attempts to gather intact tissue from extinct woolly mammoths that may be preserved in the Siberian permafrost. These scientists, Stone explains, would like to find mammoths' sex cells (sperm or eggs) with intact DNA in order to attempt to clone these animals, which have been extinct for thousands of years. Stone relates how Kazufumi Goto, the project's leader, and his entrepreneur partner, Kazutoshi Kobayashi, came to undertake their unusual quest. He also recounts some of the frustrations they have encountered in their Siberian trips and their hopes of building a "Pleistocene Park" to house their cloned mammoths and other extinct animals.

"Look, over here! Hair!" shouts one of the sperm hunters, pointing to a frayed brown tangle protruding from a cliff along the Kolyma River in Siberia. The young man tugs gently at the strands, in the hope that they're attached to a hulk that long ago lay down for the last time in this Ice Age sediment.

Ecologist Sergei Zimov, a big-boned Russian with a thick blond beard, slogs through the mud along the river toward the find. Behind him, on the horizon, a row of larch trees lean drunkenly, unable to keep a secure grip in the thin, thawed soil. It's late August, and snowflakes fluttering from the drab sky warn of an early winter. Time is running out.

As soon as Zimov gets close enough to touch the hair, he unceremoniously rips a few strands free and rolls them across his fingers, pondering.

Could this, finally, be the hair of a woolly mammoth, its frozen body—and more important, its genitalia—locked inside the crumbling black cliff? Could this be the first step in a bizarre quest led by Japanese biologists to inseminate an Asian elephant with woolly mammoth sperm and selectively breed a fabled prehistoric creature that became extinct thousands of years ago? "I know it sounds unbelievable," says Kazufumi Goto, the leader of the project, "but no science can deny our idea."

Zimov turns the hair in his hands, rubs it between his fingers and thumb, then poisons the air with his words: "Steppe bison." He yanks out the rest of the clump to reveal only more loess [soil], no skin or meat.

"Not So Far-Fetched"?

Many scientists are skeptical of Goto's prospects of resurrecting even one woolly mammoth, but few dismiss the plan out of hand—a remarkable sign of how far reproductive technologies have come in only a few years. It's not the technology that's stopping Goto, it's simply a matter of finding a well-preserved woolly mammoth. "The leap of faith is finding the viable sperm or oocytes [cells that become eggs]," says John Critser, scientific director of the Cryobiology Research Institute in Indianapolis. Critser, who has transplanted elephant ovarian tissue into mice and gotten the mice to produce elephant eggs, says that producing an embryo "is not so far-fetched."

The heyday of the woolly mammoth was the Pleistocene Epoch, stretching from 1.8 million years ago to the end of the last ice age 11,000 years ago. Mammoths thrived particularly well in Siberia, where dry grasslands once stretched for hundreds of miles, supporting a vibrant ecosystem of mammoths, bison, and other jumbo herbivores. They were in turn preyed on by cave lions, wolves, and sabertoothed cats. Famished after the end of the Ice Age by a diet of low-nutrient mosses, and increasingly harried by human hunters, the big grazers dwindled to extinction.

Although most mammoths left behind only their bones, in a few cases the Siberian permafrost preserved mammoth skin and muscle. Most of the cells in this tissue had degraded, but in the past decade scientists managed to rescue a few proteins and fragmented genes to compare with those in living elephants. And until Goto began to think otherwise, that was about as close as anyone thought we'd get to mammoths again.

Life from Frozen Sperm

In the early 1980s, when Goto began his career at Kagoshima University in Japan, he applied his nascent understanding of reproductive biology to his country's pursuit of the ultimate steak. "We have to compete with American beef," he says. Goto's task was to use artificial insemination methods to develop beef with more marbling. Getting sperm from well-marbled bulls was easy: he hooked them up to an artificial vagina, a tube warmed with hot water and dabbed with pheromones, and presto. The sperm could then be frozen in glycerol and kept in a freezer indefinitely. At the time, however, collecting cow eggs wasn't so simple. "I started going to the slaughterhouse," says Goto. There he could take ovaries from cows with good marbling.

One day in 1986, while peering through a microscope at sperm swarming an egg, Goto had an epiphany. "Once a sperm attaches to the egg's membrane, it stops moving," he says. After the two membranes fuse, the sperm is engulfed by the egg and the tail breaks off. "I had been taught that the sperm gets into the oocyte by its own movement," he says. "I was so surprised to discover that wasn't true." It dawned on him that the sperm is already dead by the time it is carried into the egg and delivers its genes.

Intrigued, Goto enlisted a student, Akihiro Kinoshita, to perform a series of experiments using dead sperm. Kinoshita froze and thawed mouse sperm until their membranes were torn up and they lost their metabolism: in other words, they were dead. The genes, on the other hand, were intact. "We could freeze and thaw 20 times, and the sperm DNA would never break." But when Kinoshita tried injecting the dead sperm into mouse eggs, "it failed thousands of times," Goto says.

He didn't give up. He decided mouse eggs were too fragile and moved on to more familiar terrain, instructing his student to try using bull semen and cow eggs. "Finally, one day Kinoshita says, 'Hey, we got cleavage,'" Goto recalls. The egg had begun to divide into an embryo. Skeptical, Goto looked into the microscope. He saw a four-cell cluster where there had been two cells. Kinoshita subsequently implanted the cluster in a cow. "She got pregnant on the first try," Goto says.

At a 1992 reproductive biology meeting in Fort Collins, Colorado, Goto was peppered with questions from a reporter about the implications of the discovery. Could viable sperm be recovered from the Ice Man, the 5,000-year-old mummy that had just been discovered in the Alps? Could dead sperm save endangered species? Or extinct ones, such as a mammoth?

The Start of a Mammoth Hunt

"That was the beginning," says Goto. He wrote to several Russian scientists over the next few years, asking about frozen mammoths. No one wrote back. He queried the Russian embassy in Tokyo, but they asked him paranoid questions: Was he after valuable mammoth tusks?

Then, in 1996, Goto met Kazutoshi Kobayashi, an entrepreneur adept at doing business in Russia. In December 1993, two years after the Soviet Union dissolved, Kobayashi flew to Khabarovsk, in eastern Russia, hoping to meet scientists who wanted to license their inventions. In the years since, he has parlayed his Russian connections into a thriving operation that buys technologies such as metal-cutting gas blowtorches and oxygen generators originally designed for the Russian space station *Mir*.

Kobayashi remembers vividly his first encounter with Goto, in February 1996. "He talked like a child about his dream to make a mammoth, but he was serious," he says. "My friends always thought I was the most foolish person in the world, but there before me was someone even more foolish. I told Goto we would realize the dream together."

Kobayashi used his contacts among the government officials that oversee the vast Siberian province Yakutia to put Goto in touch with Pyotr Lazarev, director of the Mammoth Museum in the capital city, Yakutsk. In August 1996, Lazarev and Goto scouted a site near Yakutsk on the Lena River. In early 1997, Kobayashi brought Lazarev to Japan to plan a full-blown expedition for that summer.

Breeding or Cloning?

Meanwhile, Goto found a scientific partner, Akira Iritani, chairman of the department of genetic engineering at the prestigious Kinki University in Japan. Iritani came with impeccable credentials: in 1986 his lab was one of the first in the world to successfully fertilize rabbit eggs employing a technique now used in humans. Iritani also has an abiding interest in saving endangered species. Kinki University has joined the Frozen Zoo project, in which several U.S. zoos swap samples of frozen semen from wild and zoo animals. Iritani's lab has sperm on ice from 35 species, including the mountain gorilla and the lion-tailed monkey. "We check on how much damage sperm can bear and still achieve fertilization," he says. "We hope we will be able to preserve these sperm to maintain a species' genetic diversity in case a problem with inbreeding develops."

Goto told Iritani his vision: find mammoth sperm with genes still intact and use it to fertilize a living elephant egg. If the resulting hybrid was a female, he could then fertilize its eggs with more mammoth sperm, breeding a line that would become more and more mammoth. Iritani didn't want to be limited to the hybrid approach. If mammoth sperm successfully fertilized an elephant egg and the resulting embryo took to a surrogate mother, a half-mammoth/half-elephant wouldn't be born for 600 days. (Elephants have the longest gestation of any mammal.) A hybrid female should reach reproductive age in 10 to 15 years. It could take 35 years, from start to finish. "I'm 15 years older than Goto," Iritani says. "I can't wait 30 years to see a mammoth."

He decided to attempt an even more improbable feat: clone a mammoth. Just a few months before, Ian Wilmut of the Roslin Institute in Scotland had stunned the world with the news that his group had cloned a sheep, Dolly, from a mammary cell of an adult sheep. "If even a small bit of mammoth tissue is found with normal DNA, it would take only a onetime procedure to clone a pure woolly mammoth," Iritani says. He believes the best place to look for intact DNA is in the somatic cells of the testicle and ovary, such as the stem cells that give rise to sperm and eggs.

Frustrations in Siberia

Goto and Iritani decided to travel to Siberia. In June 1997, Kobayashi flew to Yakutsk to obtain permits for the excavation and to ink agreements with officials regarding ownership of mammoth samples. He also took lots of cash to sponsor a preliminary dig to be led by Lazarev. But in Seoul, between planes on the way to Yakutsk, a bag packed with $60,000 disappeared. "It was terrible," Kobayashi says. "I had to go to Russia without any money."

Arriving empty-handed, he scraped together a few thousand dollars from associates and sent Lazarev's team on a 1,000-kilometer helicopter trek to Chokurdakh, a tiny settlement on the Indigirka River. There the team excavated at a nearby site for a few days, finding a chunk of mammoth leg. Although the meat was too degraded to yield intact DNA, the find suggested that Chokurdakh was the spot to return to that August. A week before the Japanese were to depart for Russia, Chokurdakh's mayor demanded a tax of $70,000 to enter his demesne. Kobayashi balked, negotiations failed, and Lazarev had to look for other places to dig.

He settled on Duvannyi Yar, a renowned mammoth site reached by flying to Cherskii, a town that once served as a base camp for gold prospectors. Lazarev phoned Sergei Zimov, director of the Northeast Scientific Station there, who agreed to help. But troubles plagued the researchers as soon as they arrived in Russia. At the Khabarovsk airport, customs officials seized a satellite phone and a detailed map of the region published by a U.S. agency. It didn't matter that Kobayashi had already secured all the necessary paperwork. "The customs officers just wanted us to pay arbitrary fines," says Iritani. Money changed hands, but the crew lost three precious days and their equipment remained impounded.

Finally, on August 14, the team hopped a flight to Yakutsk to meet Lazarev and fly to Cherskii. There they boarded another plane, chartered by Kobayashi, which to their surprise was loaded with black-market goods. After two days of arguing and haggling, Kobayashi paid the pilot, and the crew squeezed on the plane. When they finally got to Cherskii, the team was relieved to find Zimov and his staff ready for them. He had managed to buy a barge on which he built a shelter for sleeping and cooking. "After all the troubles we had, the houseboat looked like a luxury hotel," says Kobayashi.

Scouring the Cliffs

The expedition anchored near Duvannyi Yar and forayed to shore by rubber boat. The research station's little black poodle led the way across the mucky beach. "He was like a scout, showing us the dangerous places where we shouldn't step," says Iritani.

The scientists set to work scouring the cliffs for any sign of mammoth flesh. Zimov has found that these frozen Pleistocene sediments span 90,000 years and contain about 100 mammoth skeletons per square kilometer. From this density, he estimates that every 10 square kilometers of Pleistocene steppe could support one living mammoth. How many are anywhere close to intact is anybody's guess: it all depends on how they died.

For Goto and his colleagues, an ideal scenario would be for a mammoth to have plunged into an ice crevasse in winter and have frozen rapidly. Another favorable way would be for a mammoth to have drowned in a small lake and settled into the permafrost at the bottom.

Using picks to probe the sediments at the cliff face, the researchers homed in on spots with mammoth bones or swatches of hair. They scampered up and down cliffs, avoiding the occasional curtains of ice. By the third day, all the expedition had to show for its efforts was the discovery of an ancient horse skeleton. They couldn't even get that out of the cliff, because it was too high for their water hoses to blast it free. The weather started to turn foul, and the Russians pressed to return to Cherskii. "It was frustrating not to come back with tissue," Kobayashi says.

The team has since encountered more obstacles. In August 1998 they hoped to visit the Taimyr Peninsula, but the price of chartering a helicopter had gone up to $500 an hour—ten times more than the previous year; Goto couldn't travel because his son was ill, and Iritani was booked for several conferences abroad. Zimov's team kept an eye out for mammoth tissue while doing their annual climate-change research at Duvannyi Yar. They found plenty of bison hair and mammoth bones, but no sperm.

The Japanese haven't given up. In 1999 they plan to go to one of several sites where there have been reports of frozen mammoths. Iritani says his university has spread the word that it will pay 1 million yen ($9,000) for mammoth tissue preserved well enough for experiments, a bounty he hopes will entice Russians who illegally collect mammoth tusks for ivory.

Pleistocene Park

If Goto and his colleagues do find the mammoth of their dreams, they are prepared to spare no expense. They'll return to Japan and bring back a liquid nitrogen container in which to preserve the cells, along with enough money to book the necessary planes and helicopters.

And if the researchers succeed, the new home of their living dead could be Pleistocene Park, a 160-square-kilometer preserve near Duvannyi Yar. In 1998, Zimov and several colleagues reintroduced 32 Yakutian horses into the fenced-off area, the first phase of an experiment to test whether the horses, along with moose, reindeer, and a herd of bison imported from Canada, will rip up the moss and shrubs with their hooves and teeth, allowing the grasses that dominated the region in the Ice Age to return.

The Siberian steppes might then return to their former ecological glory. Zimov says: "I hope the density of animals in the park in 20 years will be the same as in the Serengeti." The cold climate and resurgent grasses would be a perfect home for resurrected mammoths. But Zimov points out that they would face bleak odds of long-term survival unless they have enough genetic variation to avoid inbreeding. "We would have to have a minimum of 10 to 20 variations in the mammoth genome for the herd to survive."

Goto would be happy with even a single mammoth, for it would fulfill a pledge he made to children in his hometown. Early in 1998, he made a drastic career change, giving up his position at Kagoshima University to head the Ezu Kindergarten, which his family owns, in Kumamoto. Goto says he has always wanted to help shape the minds of children long before they get to a university. "Some people say, 'The mammoth is dead under the ice. Why don't you leave it in peace?'" Goto says, "But I promised the kids I would create a mammoth. I want to do my best to accomplish the children's dream."

CHAPTER 3

CLONING AND
THE MEDIA

CLONING FACTS AND FICTIONS

Jon Turney

In the following selection, Jon Turney describes how fictional portrayals of cloning and other forms of human gene alteration shape the public's response to actual scientific developments in the field. Public confusion has been heightened, Turney believes, by writings such as *In His Image*, a 1978 book in which American science journalist David Rorvik claimed that he had assisted in the cloning of a human being. Rorvik's book blurred the line between fact and fiction, Turney asserts, just as imaginary scenarios of human cloning spawned by the factual reports of the successful 1997 cloning of a sheep have done. On the other hand, Turney points out that fiction can also be a useful way of framing ethical questions raised by scientific research into controversial areas such as cloning. Turney is the author of *Frankenstein's Footsteps: Science, Genetics, and Popular Culture*, from which the following is excerpted. He is also a senior lecturer in science communication in the department of science and technology studies at University College, London.

Cloning has long been one of the possibilities used to symbolise the powers of new biological technology. To some extent, this began with Aldous Huxley's vision in *Brave New World*, but it really emerged as a recurrent motif in debate in the late 1960s and early 1970s. The contention over fact and fiction was then crystallised by a book published by the US science journalist David Rorvik in 1978. In the book, *In His Image*, Rorvik claimed he had assisted in the cloning of a human being. He presented his story as fact, and provided extensive references to back up his claim that such a feat was possible. The claim was strongly denied by scientists, most notably by the British biologist Derek Bromhall, who was angry that his own work had been cited, and provided a convincing demolition of the scientific credibility of the story. Bromhall also accused Rorvik of presenting fiction as fact out of greed, glossing over his own claim about his motives in his epilogue where he expressed the hope that many readers will be persuaded of the possibility, perhaps even the probability of what I have described and benefit by this 'preview' of an astonishing development whose time . . . has apparently not yet quite come. And if this book, for whatever reason, increases public interest and participation in decisions related to genetic engineering then I will be more than rewarded for my efforts.

Here, Rorvik seemed to admit that the book was an attempt to exploit the difficulties scientists have in responding to claims like his. By deliberately blurring the boundary between fact and fiction, he hoped to provoke public discussion. In that he succeeded, aided considerably by the welter of scientific denunciation. Although the book was condemned as a literary confidence trick, it was widely read. What was, in truth, a badly written novel, with copious discussion of bioethics culled from the academic literature, became a Literary Guild selection, and the US paperback rights were sold for a quarter of a million dollars.

In addition, Rorvik's earlier non-fiction collection on the biological revolution was republished in paperback, and the controversy spawned numerous newspaper and magazine articles, at least one conventional popular non-fiction book on cloning and a British television documentary. Although this programme generally supported the view that the book was a hoax, the preview article in the *Radio Times,* replete with references to Huxley, concluded that "in one sense . . . it doesn't matter if it is true or not . . . as the chronology demonstrates, 'if it be not now, yet it will come. The readiness is all'. We *live* in a brave new world."

With this kind of response, Rorvik seemed justified in believing that his deliberate blurring of genre boundaries would be effective in stimulating debate. Other writers at the time were pursuing the issues [raised by biotechnology] in fiction, but not getting this kind of attention. Consider, for example, a science fiction novel of the near future by the well-regarded Australian author George Turner, who has a character observe:

> As a group, biologists are the most dangerous men alive. The bomb we've learned to live with and pollution we will handle. But biologists!

> What they have achieved since the sixties is enough to put the fear of hellfire into Jehovah himself. Artificial inovulation, the gerontological drugs, brain regrowth and the mechanics of gene manipulation—these are already with us, imperfect and unready but with us.

> They are only the beginning.

> Consider the implications, and retch.

This is a rather stronger condemnation of contemporary biological research than anything offered by Rorvik, but it is the latter's book which is still remembered, at least when journalists look up their cuttings to write another article on cloning.

Using Fictions to Frame a Debate

The succession of bursts of publicity about cloning show how different parties to the debates about the future implications of biological science treat the role of fictions in these debates in different ways. Some commentators explicitly acknowledge the importance of a fictional tradition in offering symbolic possibilities which dramatise a wider range of concerns about experimental biology. Consider, for example, one notable later media flurry about cloning, when Jerry Hall and colleagues at George Washington University in the US made the front page of the *New York Times* in 1993 with experiments aimed at producing multiple embryos from a single *in vitro* fertilised egg. The lengthy reports which followed in *Time* and *Newsweek* both emphasised the distance between what Hall and colleagues had achieved and the many fictional uses of cloning. Under the heading 'Clone Hype', *Newsweek*'s main report made continual references both to Rorvik's book and to stories about cloned dinosaurs and multiple Hitlers to stress what the work was *not* about. *Time* took a very similar line, while emphasising that the work had created a worldwide sensation, and that 77 per cent of Americans polled that week wanted to see such research either temporarily halted or banned outright. Their reporter observed that the actual research reported 'seems, in many ways, unworthy of the hoopla.' And the magazine made a point of showing how many of the bioethicists who contributed comment to the wider press coverage of the story were offering scenarios that went well beyond existing technology. This suggestion was underlined with a separate article on 'Cloning classics', summarising fictions about the subject from *Brave New World* to Fay Weldon's 1989 novel *The Cloning of Joanna May*. The article opened with the observation that, 'When it comes to dealing with cloning, ethicists and science-fiction writers have almost identical job descriptions.'

The outpouring of concerned commentary which followed Hall's announcement was relatively short-lived. But the most recent episode in the cloning saga evoked a longer-running debate. The cover of *Nature* for 27 February 1997 depicted a Scottish-bred sheep, known as Dolly, over the headline 'A flock of clones'. By the time *Nature* appeared, the story had already been widely previewed in the general media, following a British Sunday newspaper's exclusive the previous weekend. The birth of Dolly was taken as a signal that cloning of adult mammals, and hence humans, was now a real technological possibility.

A team at the Roslin Institute in Edinburgh, working with colleagues from a local biotechnology firm, had perfected techniques

which allowed nuclei from adult cells to be fused with eggs to produce a new individual genetically identical to the original adult. Their
success rate was very low, one lamb from 277 fusions, but they did
succeed. The enormous volume of comment that followed . . . [had]
two particularly prominent features . . . which I want to highlight.
One was the apparently strongly felt urge, both among media writers
and, as opinion polls had already suggested, much of the public, that
something must be done to stop human cloning. The other was, as
Nature's own editorial argued, that the widespread feeling that governments had been 'caught napping by clones' was an indictment of
all the policy-makers, ethical advisers and technology foresight panels
who are supposed to be keeping an eye on science and technology.
This shows, perhaps, one ultimate limitation of the urge to dismiss
particular possibilities which have become the focus of concern about
the direction biology may be taking as fictions. The paradoxical result
in Dolly's case was that she was the realisation of an idea which had
been discussed for more than half a century, yet her advent in the
flesh was treated as an enormous surprise. *Newsweek* suggested that

> Twenty years ago, when only the lowly tadpole had been
> cloned, bioethicists raised the possibility that scientists might
> some day advance the technology to include human beings as
> well. They wanted the issue discussed. But scientists assailed
> the moralists' concerns as alarmist. Let the research go for-
> ward, the scientists argued, because cloning human beings
> would serve no discernible scientific purpose. Now the cloning
> of humans is within reach, and society as a whole is caught
> with its ethical pants down.

If so, it was not for want of trying on the part of some commentators.
But this time, although there were still respectable arguments why
human cloning was unlikely ever to be possible—which some scientists were quick to point out—there was a widely shared determination that the possibility should be considered seriously. Those who
responded to Dolly, from President Bill Clinton on down, the former
with his call for a report 'within 90 days' from his advisers, wanted to
understand more clearly what human cloning might mean.

How the Public Understands Fictions

Although it was widely recognised that cloning was still mainly a
symbol for a broader set of technologies, that there were important
issues relating to the industrial use of animals connected with Dolly's
future as a drug incubator, and that cloned humans would probably
not be identical to their genetic forebears, it was the prospect of an
end to biological individuality which really caught media audiences'
attention. Fictional scenarios abounded as writers tried to help readers
think through the possible implications. *Time* magazine, for example,

which like *Newsweek* again made cloning its cover story, ran a four-page article on future ethical problems built around a series of tableaux from a hospital cloning laboratory of the future. To top that, it rounded off its special report on Dolly with a tongue-in-cheek science fiction story by Douglas Coupland. The magazine seemed to be taking seriously its suggestion in 1993 that ethicists and science fiction writers had similar jobs.

This seems to suggest that creating fictions about the possible outcomes of applying biological technology to people is a legitimate contribution to debate. Both literary creation and scenario-spinning by bioethicists are ways of alerting society to possibilities which merit discussion before they are realised—certainly a view writers tend to share. Scientists, though, are not so sure. One stance they adopt more commonly is to argue that the public is unable to distinguish fact from fiction. Non-scientists, it is suggested, interpret metaphorical warnings literally. Writers must therefore take responsibility for portraying well-intentioned science in a negative light. Commenting on films in the *Frankenstein* tradition, the distinguished British geneticist Paul Nurse suggests that 'the real dilemma comes when the freedom of the artist to produce what they like has to be combined with the fact that these productions are taken by the public to be an absolutely true portrayal of science'. His British colleague Ruth McKernan agreed that 'the line between science fantasy as entertainment and science fact needs to be drawn more clearly'.

This kind of assertion radically oversimplifies the relations between media and audiences, fact and fiction, and the range of stories available at any one time. 'Factual' stories are always framed in some way which is intended as a guide to interpretation—one reason for the journalistic invocations of *Frankenstein* or *Brave New World*. But this does not mean that the suggestion that some piece of science is 'like' Frankenstein's project or 'reminiscent' of *Brave New World* is taken literally. Both literary criticism and media studies demand that we take a more sophisticated view of what goes on when diverse audiences assimilate a complex set of messages about science.

"CLONING" MOVIE SCRIPTS

Lisa Bannon and Frederick Rose

In the humorous article written shortly after the successful cloning of a lamb was announced in February 1997, *Wall Street Journal* staff writers Lisa Bannon and Frederick Rose report that one of the first effects of the news was an increase in Hollywood producers' interest in movies about cloning. Previous movies concerning cloning, such as Michael Keaton's 1996 release *Multiplicity*, have often been box-office failures, they write. However, the authors reveal, new public interest in cloning has led movie moguls to reconsider projects with this theme that had been put on hold. Bannon and Rose mention several upcoming films that have plots partly or completely focused on cloning. They also describe frenzied meetings of writers preparing cloning jokes and skits for television comedy and talk shows. They point out, however, that some producers—not to mention the scientists associated with actual cloning research—have doubts about the wisdom of concentrating on the science-fiction aspects of cloning.

For Lewis Kleinberg, the news in February 1997 that Scottish scientists had cloned an adult sheep was disturbing. In fact, he didn't sleep for 50 hours after hearing about it. That is because Mr. Kleinberg, a screenwriter, and his producer partner Stephen Yakobian have spent the past two months working on a script outline for an action thriller about cloning called *The Seed*.

"Suddenly, it's splashed all over the newspaper! Even the scenes I had in my head are being played out on TV all over the world!" Mr. Kleinberg says breathlessly from his car phone as he coasts along Santa Monica Boulevard.

"We planned to wait and take it around the studios later, but hype is big business in Hollywood," he says. "We're trying to finish it up today. You know 50,000 other people are contacting the scientists in Scotland. The sense of panic is setting in."

The scientific breakthrough might turn out to be important in medicine, agriculture and even theology. But it already has had a broad impact in one industry: entertainment. Scripts are being dusted off. All of Hollywood is agog. And, for comedians—well, just as O.J. Simpson jokes are beginning to fade, think of the discovery as a kind of full-employment act for America's gag writers.

Jay Leno devoted the better part of his monologue to the cloning story, including a segment in which he introduced triplets as the Scottish researchers who conducted the experiment.

"It has all the comedy workings right there: science and sheep," says Eddie Feldmann, head writer for HBO's *Dennis Miller Live*. Mr. Feldmann says it will be a challenge, however, to come up with a "fresh angle.". . .

The writers at *Politically Incorrect* spent the better part of a day in and out of sheep-joke meetings. They were scrambling to find jokes that would set host Bill Maher apart from the flock.

"We don't quite have it yet," frets Chris Kelly, supervising writer for the late-night ABC show. "We have to sit down with Bill and thrash it out some more. We're working on the Scotland angle. We're thinking the Scots are too cheap to buy an acrylic sweater," Mr. Kelly jokes, testing his one-liner. "Viewers are faxing us jokes about virgin wool."

Resurrecting Scripts

Around Hollywood, producers are busily trying to resurrect dead scripts about cloning. Phillip Goldfine, a production executive at Trimark Pictures, came back from lunch with an agent thinking he might buy a four-year-old script about a clone with a computer-generated past. The project, originally called *Shatter Game* but now renamed *Created Equal,* had been in development at two TV studios before being shelved in 1995 because of disinterest.

But Mr. Goldfine was too late. Before he could make a commitment on the script, a production company had snatched it from under his nose, hoping to sell it to a major network. The same agent who hawked *Created Equal,* Anne McDermott, has two other clone scripts that she couldn't sell before. "But with this new sheep situation, I might have some luck." She adds: "It's the sheep following the sheep."

Hollywood, of course, invented cloning—and not just of its own storylines. "Anybody who thinks Hollywood doesn't have tremendous cloning skills hasn't seen the *Die Hard* movies," says Richard Jeni, who has incorporated the cloning news into his routine at the Laugh Factory comedy club in Hollywood.

Long before researchers at the Roslin Institute in Edinburgh created a ewe named Dolly, movies and TV series had exploited both the scary and the funny side of duplicating nature. In one of the best known, the 1978 film *The Boys from Brazil,* Gregory Peck plays an evil Nazi doctor in hiding who clones Adolf Hitler in an attempt to reconquer the world.

But most studios had abandoned cloning scripts after the movie *Multiplicity,* starring Michael Keaton as a harried husband who clones himself to meet the demands of work and family, fared poorly at the box office in July 1996.

"If the sheep had been cloned a year earlier, who knows what would have happened to my movie," Mr. Keaton says ruefully.

Now, even the most bizarre ideas have currency. "I've just read a script for a TV movie about a nurse who finds that an accident victim brought into a hospital E.R. is a clone of herself, and she goes in search of the woman's identity," says Beth Schroeder, vice president of development at O'Hara-Horowitz Productions, who produced the TV movie *A Child's Wish* for CBS in January 1997 featuring a cameo appearance by President Bill Clinton. "I have a feeling this script has been around for a while," she adds, "but this gives me more impetus to go after it."

Dolly the lamb has turned out to be a woolly coincidence for Howard Braunstein, a partner in the production company Jaffe/Braunstein Films. Mr. Braunstein is delivering to CBS a script for a four-hour mini-series based on Ken Follett's current bestseller about cloning, *The Third Twin*.

In a year in which two volcano movies, two asteroid movies and a cluster of other disaster flicks are about to hit theater and TV screens around the world, cloning may bring some scientific relief to Hollywood. "Disasters are dead," says Rob Carlson, a literary agent at the William Morris Agency, who predicts that many of the cloning movies that have been languishing in Hollywood's version of purgatory—"development hell"—will now get made.

Some producers have already been seeing double-helix signs. The king of B movies, Roger Corman, plans to complete a cloning movie called *Future Fear* in March 1997 with a likely release on Showtime cable network. Starring Stacy Keach, it involves mixing the cells of humans and chimpanzees to create clones of a new species to fight a deadly virus from outer space.

Sony's Columbia Pictures has a film set for release in the fall of 1997 called *Gattaca*, starring Uma Thurman and Ethan Hawke, about what happens when people can control their own evolutions. And in the fall of 1997 Twentieth Century Fox will release the fourth in its *Alien* series of films. For those who thought this saga died with Sigourney Weaver's Ripley character in *Alien 3*, guess again. Ripley has been cloned and returns in *Alien Resurrection*.

The Clones Stop Here

Not everyone in Hollywood thinks that all of these ideas are worth cloning, however. Joe Roth, chairman of Walt Disney Studios, says simply: "I've never in my life read a cloning script I wanted to make."

As for those Scottish scientists, they may not yet know what they have let themselves in for. Dr. Harry Griffin, assistant director of the Roslin Institute, says he doesn't know whether any Hollywood offers have come in yet because his office is so overwhelmed by the worldwide press attention.

That said, Dr. Griffin says he has been rattled by some of the initial press coverage. The focus on the sci-fi, human cloning aspects of the experiments has made him wary about hooking up with TV or movie producers.

"There have been some quite remarkable extrapolations," he says. "I don't think it serves any useful purpose, feeding people's fantasies."

In the end, he says his job is to protect the reputation of what is basically a scientific institute. "Somebody has to draw the line," Dr. Griffin says. "Our work is in sheep."

THE MESSAGES OF *JURASSIC PARK*

Raymond G. Bohlin

In the following article, Raymond G. Bohlin describes what he claims are the philosophical messages behind the popular 1993 film *Jurassic Park*, which depicts scientists cloning dinosaurs from bits of DNA extracted from amber. According to Bohlin, both the film and the Michael Crichton novel on which it is based accuse scientists of making unwarranted intrusions on nature. At the same time, he contends, the movie presents assumptions about an evolutionary relationship between dinosaurs and birds that, in Bohlin's opinion, are not supported by scientific evidence. Finally, he points out the unlikelihood of the movie's basic premise: that long-extinct animals can be cloned from small amounts of DNA. Bohlin is a molecular biologist and population geneticist as well as the director of communications at Probe Ministries, whose mission is to reclaim the primacy of Christian thought and values in Western culture. He is also the coauthor of *The Natural Limits to Biological Change*.

Driving home after seeing the movie *Jurassic Park* in the first week of its release in 1993, I kept seeing tyrannosaurs and velociraptors coming out from behind buildings, through intersections, and down the street, headed straight at me. I would imagine: What would I do? Where would I turn? I certainly wouldn't shine any lights out of my car or scream. Dead give-aways to a hungry, angry dinosaur. Then I would force myself to realize that it was just a movie. It was not reality. My relief would take hold only briefly until the next intersection or big building.

In case you can't tell, I scare easily at movies. *Jurassic Park* terrified me. It all looked so real. Steven Spielberg turned out the biggest money-making film in history. Much of the reason for that was the realistic portrayal of the dinosaurs. But there was more to *Jurassic Park* than great special effects. It was based on the riveting novel by Michael Crichton and while many left the movie dazzled by the dinosaurs, others were leaving with questions and new views of science and nature.

Excerpted from "The World View of *Jurassic Park*," by Raymond G. Bohlin, Probe Ministries, April 1997. Reprinted with permission from the author. Article available at www.probe.org/docs.Jurassic.html.

The movie *Jurassic Park* was terrific entertainment, but it was entertainment with a purpose. The purpose was many-fold and the message was interspersed throughout the movie, and more so throughout the book. My purpose in this essay is to give you some insight into the battle that was waged for your mind throughout the course of this movie.

Jurassic Park was intended to warn the general public concerning the inherent dangers of biotechnology first of all, but also science in general. Consider this comment from the author Michael Crichton:

> Biotechnology and genetic engineering are very powerful. The film suggests that [science's] control of nature is elusive. And just as war is too important to leave to the generals, science is too important to leave to scientists. Everyone needs to be attentive.

Overall, I would agree with Crichton. All too often, scientists purposefully refrain from asking ethical questions concerning their work in the interest of the pursuit of science.

But now consider director Steven Spielberg, quoted in the pages of the *Wall Street Journal:* "There's a big moral question in this story. DNA cloning may be viable, but is it acceptable?" And again in the *New York Times*, Spielberg said, "Science is intrusive. I wouldn't ban molecular biology altogether, because it's useful in finding cures for AIDS, cancer and other diseases. But it's also dangerous and that's the theme of *Jurassic Park*." So Spielberg openly states that the real theme of *Jurassic Park* is that science is intrusive.

In case you are skeptical of a movie's ability to communicate this message to young people today, listen to this comment from an eleven-year-old after seeing the movie. She said, "*Jurassic Park*'s message is important! We shouldn't fool around with nature." The media, movies and music in particular, are powerful voices to our young people today. We cannot underestimate the power of the media, especially in the form of a blockbuster like *Jurassic Park*, to change the way we perceive the world around us.

Many issues of today were addressed in the movie. Biotechnology, science, evolution, feminism, and New Age philosophy all found a spokesman in *Jurassic Park*.

The Dangers of Science

The movie *Jurassic Park* directly attacked the scientific establishment. Throughout the movie, Ian Malcolm voiced the concerns about the direction and nature of science. You may remember the scene around the lunch table just after the group has watched the three velociraptors devour an entire cow in only a few minutes. Ian Malcolm brashly takes center stage with comments like this: "The scientific power . . . didn't require any discipline to attain it. . . . So you don't take any responsibility for it." The key word here is responsibility. Malcolm intimates that Jurassic Park scientists have behaved irrationally and irresponsibly.

Later in the same scene, Malcolm adds, "Genetic power is the most awesome force the planet's ever seen, but, you wield it like a kid that's found his dad's gun." Genetic engineering rises above nuclear and chemical or computer technology because of its ability to restructure the very molecular heart of living creatures. Even to create new organisms. Use of such power requires wisdom and patience. Malcolm punctuates his criticism in the same scene when he says, "Your scientists were so preoccupied with whether or not they could, they didn't stop to think if they should."

Malcolm's criticisms should hit a raw nerve in the scientific community. As Christians we ask similar questions and raise similar concerns when scientists want to harvest fetal tissue for research purposes or experiment with human embryos. If Malcolm had limited his remarks to Jurassic Park only, I would have no complaint. But Malcolm extends the problem to science as a whole when he comments that scientific discovery is the rape of the natural world. Many youngsters will form the opinion that all scientists are to be distrusted. A meaningful point has been lost because it was wielded with the surgical precision of a baseball bat.

Surprisingly, computers take a more subtle slap in the face—surprising because computers were essential in creating many of the dinosaur action scenes that simply could not be done with robotic models. You may remember early in the movie, the paleontological camp of Drs. Grant and Satler where Grant openly shows his distrust of computers. The scene appears a little comical as the field-tested veteran expresses his hate for computers and senses that computers will take the fun out of his quaint profession.

Not so comical is the portrayal of Dennis Nedry, the computer genius behind Jurassic Park. You get left with the impression that computers are not for normal people and the only ones who profit by them or understand them are people who are not to be trusted. Nedry was clearly presented as a dangerous person because of his combination of computer wizardry and his resentment of those who don't understand him or computers. Yet at the end of the movie, a young girl's computer hacking ability saves the day by bringing the system back on line.

The point to be made is that technology is not the villain. Fire is used for both good and evil purposes, but no one is calling for fire to be banned. It is the world view of the culture that determines how computers, biotechnology, or any other technology is to be used. The problem with Jurassic Park was the arrogance of human will and lack of humility before God, not technology.

There were many obvious naturalistic or evolutionary assumptions built into the story which, while not totally unexpected, were too frequently exaggerated and overplayed.

For instance, by the end of the book and the film you felt bludgeoned by the connection between birds and dinosaurs. Some of these connections made some sense. An example would be the similarities between the eating behavior of birds of prey and the tyrannosaur. It is likely that both held their prey down with their claws or talons and tore pieces of flesh off with their jaws or beaks. A non-evolutionary interpretation is simply that similarity in structure indicates a similarity in function. An ancestral relationship is not necessary.

But many of the links had no basis in reality and were badly reasoned speculations. The owl-like hoots of the poison-spitting dilophosaur jumped out as an example of pure fantasy. There is no way to guess or estimate the vocalization behavior from a fossilized skeleton.

Another example came in the scene when Dr. Alan Grant and the two kids, Tim and Lex, meet a herd of gallimimus, a dinosaur similar in appearance to an oversized ostrich. Grant remarks that the herd turns in unison like a flock of birds avoiding a predator. Well, sure, flocks of birds do behave this way, but so do herds of grazing mammals and schools of fish. So observing this behavior in dinosaurs no more links them to birds than the webbed feet and flattened bill of the Australian platypus links it to ducks! Even in an evolutionary scheme, most of the behaviors unique to birds would have evolved after the time of the dinosaurs.

A contradiction to the hypothesis that birds evolved from dinosaurs is the portrayal of the velociraptors hunting in packs. Mammals behave this way, as do some fishes such as the sharks, but I am not aware of any birds or reptiles that do. The concealment of this contradiction exposes the sensational intent of the story. It is used primarily to enhance the story, but many will assume that it is a realistic evolutionary connection.

Finally, a complex and fascinating piece of dialogue in the movie mixed together an attack on creationism, an exaltation of humanism and atheism, and a touch of feminist male bashing. I suspect that it was included in order to add a little humor and to keep aspects of political correctness in our collective consciousness. Shortly after the tour of the park begins and before they have seen any dinosaurs, Ian Malcolm reflects on the irony of what Jurassic Park has accomplished. He muses, "God creates dinosaurs. God destroys dinosaurs. God creates man. Man destroys God. Man creates dinosaurs." To which Ellie Satler replies, "Dinosaurs eat man. Woman inherits the earth!" Malcolm clearly mocks God by indicating that not only does man declare God irrelevant, but also proceeds to duplicate God's creative capability by creating dinosaurs all over again. We are as smart and as powerful as we once thought God to be. God is no longer needed.

While the movie was not openly hostile to religious views, Crichton clearly intended to marginalize theistic views of origins with humor, sarcasm, and an overload of evolutionary interpretations.

Jurassic Park and the New Age

Ian Malcolm, in the scene in the biology lab as the group inspects a newly hatching velociraptor, pontificates that "evolution" has taught us that life will not be limited or extinguished. "If there is one thing the history of evolution has taught us, it's that life will not be contained. Life breaks free. It expands to new territories, it crashes through barriers, painfully, maybe even dangerously, but, uh, well, there it is! . . . I'm simply saying that, uh, life finds a way."

Evolution is given an intelligence all its own! Life finds a way. There is an almost personal quality given to living things, particularly to the process of evolution. Most evolutionary scientists would not put it this way. To them evolution proceeds blindly, without purpose, without direction. This intelligence or purposefulness in nature actually reflects a pantheistic or New Age perspective on the biological world.

The pantheist believes that all is one and therefore all is god. God is impersonal rather than personal and god's intelligence permeates all of nature. Therefore the universe is intelligent and purposeful. Consequently a reverence for nature develops instead of reverence for God. In the lunchroom scene Malcolm says, "The lack of humility before nature being displayed here, staggers me." Malcolm speaks of Nature with a capital "N." While we should respect and cherish all of nature as being God's creation, humility seems inappropriate. Later in the same scene, Malcolm again ascribes a personal quality to nature when he says, "What's so great about discovery? It's a violent penetrative act that scars what it explores. What you call discovery, I call the rape of the natural world." Apparently, any scientific discovery intrudes upon the private domain of nature. Not only is this New Age in its tone, but it also criticizes Western culture's attempts to understand the natural world through science.

There were other unusual New Age perspectives displayed by other characters. Paleobotanist Ellie Satler displayed an uncharacteristically unscientific and feminine, or was it New Age, perspective when she chastened John Hammond for thinking that there was a rational solution to the breakdowns in the park. You may remember the scene in the dining hall, where philanthropist John Hammond and Dr. Satler are eating ice cream while tyrannosaurs and velociraptors are loose in the park with Dr. Grant, Ian Malcolm, and Hammond's grandchildren. At one point, Satler says, "You can't think your way out of this one, John. You have to feel it." Somehow, the solution to the problem is to be found in gaining perspective through your emotions, perhaps getting in touch with the "force" that permeates everything around us as in *Star Wars*.

Finally, in this same scene, John Hammond provides a rather humanistic perspective on scientific discovery. He is responding to Ellie Satler's criticisms that a purely safe and enjoyable Jurassic Park is not possible. Believing that man can accomplish anything he sets his mind to, Hammond blurts out, "Creation is a sheer act of will!" If men and women were gods in the pantheistic sense, perhaps this would be true of humans. But if you think about it, this statement is truer than first appears, for the true Creator of the universe simply spoke and it came into being. The beginning of each day's activity in Genesis 1 begins with the phrase, "And God said."

Creation is an act of will, but it is the Divine Will of the Supreme Sovereign of the universe. And we know this because the Bible tells us so!

"They Clone Dinosaurs, Don't They?"

The movie *Jurassic Park* raised the possibility of cloning dinosaurs. Prior to the release of the movie, magazines and newspapers were filled with speculations concerning the real possibility of cloning dinosaurs. The specter of cloning dinosaurs was left too much in the realm of the eminently possible. Much of this confidence stemmed from statements from Michael Crichton, the author of the book, and producer Steven Spielberg.

Scientists are very reluctant to use the word "never." But this issue is as safe as they come. Dinosaurs will never be cloned. The positive votes come mainly from Crichton, Spielberg, and the public. Reflecting back on his early research for the book, Michael Crichton said, "I began to think it really could happen." The official *Jurassic Park* Souvenir magazine fueled the speculation when it said, "The story of *Jurassic Park* is not far-fetched. It is based on actual, ongoing genetic and paleontologic research. In the words of Steven Spielberg: This is not science fiction; it's science eventuality." No doubt spurred on by such grandiose statements, 58% of 1000 people polled for *USA Today* said they believe that scientists will be able to recreate animals through genetic engineering.

Now contrast this optimism with the more sobering statements from scientists. The *Dallas Morning News* said, "You're not likely to see Tyrannosaurus Rex in the Dallas Zoo anytime soon. Scientists say that reconstituting any creature from its DNA simply won't work." And *Newsweek* summarized the huge obstacles when it said, "Researchers have not found an amber-trapped insect containing dinosaur blood. They have no guarantee that the cells in the blood, and the DNA in the cells, will be preserved intact. They don't know how to splice the DNA into a meaningful blueprint, or fill the gaps with DNA from living creatures. And they don't have an embryo cell to use as a vehicle for cloning." These are major obstacles. Let's look at them one at a time.

First, insects in amber. DNA has been extracted from insects encased in amber from deposits as old as 120 million years. Amber does preserve biological tissues very well. But only very small fragments of a few individual genes were obtained. The cloning of gene fragments is a far cry from cloning an entire genome. Without the entire intact genome, organized into the proper sequence and divided into chromosomes, it is virtually impossible to reconstruct an organism from gene fragments.

Second, filling in the gaps. The genetic engineers of *Jurassic Park* used frog DNA to shore up the missing stretches of the cloned dinosaur DNA. But this is primarily a plot device to allow for the possibility of amphibian environmentally-induced sex change. An evolutionary scientist would have used reptilian or bird DNA, which would be expected to have a higher degree of compatibility. It is also very far-fetched that an integrated set of genes to perform gender switching, which does occur in some amphibians, could actually be inserted accidentally and be functional.

Third, a viable dinosaur egg. The idea of placing the dinosaur genetic material into crocodile or ostrich eggs is preposterous. You would need a real dinosaur egg of the same species as the DNA. Unfortunately, there are no such eggs left. And we can't recreate one without a model to copy. So don't get your hopes up. There will never be a real *Jurassic Park*!

Ian Wilmut: A Shy Scientist Weathers a Storm of Publicity

Robert Lee Hotz

Ian Wilmut is the chief scientist of the research team at the Roslin Institute in Scotland that cloned a lamb, Dolly, from an adult body cell. In the following selection, Robert Lee Hotz describes Wilmut's reaction to the media furor that resulted after the February 1997 announcement of Dolly's existence. Hotz portrays Wilmut as a quiet person who would prefer to continue his research rather than deal with the media. However, he comments that Wilmut is quickly learning to handle the publicity that his groundbreaking work has thrust upon him. Hotz writes that Wilmut and other researchers at the institute are dismayed, although not totally surprised, that news accounts have often characterized them as irresponsible or "soulless" scientists who are unconcerned with the ethical implications of cloning. The media have also focused on the possibility of cloning humans, Hotz relates, even though the institute is doing no work in this area and Wilmut himself opposes human cloning. Hotz is a staff writer for the *Los Angeles Times* and the author of *Designs on Life: Exploring the New Frontiers of Human Fertility.*

The calm center of a storm over the science and morality of cloning is a reticent, red-whiskered Englishman named Ian Wilmut who, with the patience of one who has raised three children, is trying to reassure a world frightened by the shadows in his experiments.

In the days after the publication of his groundbreaking research, it seemed that Wilmut and his colleagues at the Roslin Institute outside Edinburgh, Scotland, had cloned not a white-faced sheep named Dolly, but a series of intractable moral dilemmas. His creation disturbed a natural order of things, opening a way into a world in which, someday, a parent and child could be identical twins.

As the man who made the sheep that shook the world, Wilmut created difficulties for himself as well, disturbing the rural tranquillity that he seems to value above almost all else.

"I am a diffident Englishman," he said. "I am trying my best to get [cloning] launched without getting my own life all tangled up in it."

But tangled up it is. "He is going to have to live with being the human face of cloning," said John Clark, one of Wilmut's senior colleagues at the institute.

A private man who has become public property, Wilmut, 52, is a modest, single-malt-sipping, Cambridge-trained rambler of the Scottish border country. He makes his home with his wife, Vivienne, and a pet spaniel in a village of barely 500 that he declines to identify.

A scientist who cares passionately about the ethics of his work, he sees himself portrayed as a kind of soulless Strangelove who heedlessly unleashed this newest threat to humanity's peace of mind.

It is an image of himself—and of science—that he and his colleagues barely recognize.

Wilmut has spent his life trying to pick the locks of the living cell, in the hope he could solve one of its most enduring mysteries. His successful cloning experiment—made public February 22, 1997—should have been the pinnacle of his career, meant to pave the way for a new field of genetic engineering.

Instead, the most concrete result of his landmark experiment may be a U.S. ban on human cloning research, to avoid something that Wilmut is emphatic should never be attempted.

"We feared the response," said Peter Sharp, who runs the institute's division of development and reproduction in which Wilmut works. "We are absolutely appalled. When things like Dolly crop up, it tests the faith in science to the limit."

The British House of Commons has requested that Wilmut testify in London on March 6; Congress wants him to answer its questions in Washington on March 12. The Vatican would like the moral meaning of cloning clarified; so would a presidential bioethics commission.

But the one man whom everyone expects to have the answers is himself bemused.

"One misconception is that scientists are any more farsighted than anyone else," Wilmut sighed. "We are not. Most of the things this technique will be used for have not yet been imagined."

Is science outracing ethics?

"I know I am struggling to keep up," Wilmut said.

Taking Heat, Sharing Credit

Wilmut has lived in Scotland for 23 years—long enough for his English accent to acquire the native burr. The institute where he works, seven miles southwest of Edinburgh, is a secluded cluster of poultry sheds, cattle barns, silos and one-story barracks of laboratories, all set off from the neighboring village of Roslin by green belts of trees, security gates and surveillance cameras.

Within this compound, about 300 molecular biologists, genetic engineers and research scientists are redesigning the livestock on which humankind depends. They are conceiving 21st-century cattle, pigs, chickens and sheep genetically tailored to produce drugs in their milk or eggs, to grow faster, fatter and be more disease-resistant.

Like Wilmut, the institute values its privacy.

Six years ago, animal rights activists set fire to the laboratories, causing several million dollars worth of damage. Today, institute officials warn visitors to avoid taking pictures that might show the license plates on staff cars.

Wilmut works on a government salary of about $60,000 a year in an office not much larger than a Land Rover. He drinks his tea from a Wallace and Gromit cartoon mug, types his research reports on a battered Macintosh II computer, and carries files home in a worn, white canvas bag.

He is an energetic wisp of a man, whose engaging gap-toothed grin was rarely in evidence.

As he prowled the institute's corridors, his colleagues had to trot to keep pace. When he sat, he coiled himself like a spring, his legs crossed and his arms folded across his chest to contain the tension. Unruly threads of reddish blond hair rose from his head like steam.

The controversy over cloning was only one of his problems.

As part of a recent economy drive, the British government cut his cloning funding. Agricultural officials maintain that the cutback had nothing to do with the furor over the cloning experiment.

Still, fumed Wilmut, "it is no way to treat people." While he certainly is the most visible person involved in the experiment, Wilmut sees himself as a scientist who succeeds by orchestrating the efforts of other researchers. He probably spends more time writing grant proposals and counseling other scientists than he does working on his personal projects.

Indeed, he attributes the key insight that made the experiment work to a colleague, Keith H.S. Campbell, who realized it was necessary to make sure that the donor cell and the egg were both in the same stage of development.

"The way you achieve things is to bring out opportunities for other people," Wilmut said. "So I don't do much experimental work. I am in lab meetings. I do some of the [animal] surgery sometimes." Others might want to be more directly involved in the experiments, but "that's the way I am," he said.

He also admitted that his hands are too clumsy for the delicate motions required in microsurgery on embryos and donor cells.

But those minor imperfections have been overshadowed by his leadership. His colleagues say he has pursued the work with remarkable tenacity for more than a decade.

"He has hit the treasure trove now, but he had a long hard time," said Alan Colman, research director at PPL Therapeutics in Roslin, a biotechnology company that collaborated in the experiment.

"Ian was in the wilderness. He was having a tough time justifying his existence."

Laying the Groundwork

Wilmut finished his doctoral degree at Cambridge University in 1973, helping to pioneer techniques to freeze sperm.

He studied in an influential but little-known animal research unit that trained a cadre of reproductive physiologists who revolutionized the science of sex. Their inventions ranged from the development of test-tube babies and the use of fertility drugs to frozen embryos and embryo surgery.

Working there, Wilmut was the first to produce a calf from a frozen embryo. There is a picture of the calf—Frosty—on his desk.

And when he left the lab, his place was taken by a young Danish scientist named Steen Willadsen who in 1986 became the first to produce multiple genetic copies of sheep by splitting cells from a developing embryo, which foreshadowed Wilmut's experiments.

"That really set the stage" for cloning, Willadsen said.

It took a decade more until Wilmut achieved what most scientists consider true cloning—growing an animal from adult cells.

By somehow prodding the genes inside an adult sheep cell into wakefulness, Wilmut and his team were able to make that cell revert to its biological infancy. From it they grew a genetic copy of the original animal from which the cell was taken.

In her guarded pen, Wilmut's brainchild peacefully chewed handfuls of Super Ewe sheep feed, pranced around and greeted the media with a full-throated bleat.

"There is this fear made flesh," said Harry Griffin, the institute's deputy science director.

In the adjacent stall, two sheep nuzzled. Meagan and Morag are clones of a slightly different variety, created by Wilmut's team in 1996 from embryonic cells. (That technique also was used by Oregon researchers, who announced in March 1997 that they had cloned two monkeys.) One of them is pregnant, proving that such cloned creatures can reproduce normally.

For the foreseeable future, however, Dolly will remain one of a kind. Wilmut said he has no immediate plans to use adult cells to clone her or any other animal.

Since her birth, however, Wilmut and his colleagues have wasted no time creating a new generation of genetically altered cloned sheep, but they are working only with clones cultivated from fetal cells. These types of experiments have greater promise for those interested in quickly producing large numbers of animals genetically engineered to produce useful pharmaceutical compounds or possess more productive traits, he said.

"Maybe we will do some more experiments with the adult cells, but we are not doing them now," he said, "because we are working toward precise genetic modifications with fetal cells."

With practiced understatement, Wilmut downplayed the importance of his work and its potential impact on humankind.

"I am not actually sure it is an incredible breakthrough. This is a useful scientific technique. It is not remotely equivalent to setting off the atomic bomb.

"There is an analogy that I think is more appropriate and that is organ transplantation," he said. "We forget what a stir that caused around the world, in part because of the emotional attachment people place on the heart."

Nonetheless, Dolly does turn some of the simplest human preconceptions on their head. Even something as rudimentary as her age is imponderable. Is she actually 8 months old—her apparent gestational age—or is she more than 6 years old—the age of the animal from whose udder cell she was cultivated?

"That's a fascinating question," Wilmut said politely. "But we don't have the answer yet."

Only a detailed analysis of genes will reveal her true biochemical age.

Irked by Nightmare Scenarios

While he shares the public concern over the technology he has created, Wilmut is irritated by some of the more exotic nightmare scenarios being outlined and is angered by the assumption that someone who has spent his life working in his field might be blind to its unsettling implications.

Wilmut says he has no religious beliefs. His wife is an elder of the Church of Scotland. He has worked closely with church leaders assessing the moral implications of such research. When it came time to patent the cloning technique in 1997, he insisted that the patent language explicitly exclude any application to humans.

His hands fidgeted as he talked, clasping and unclasping. It was the only sign that a man under rigid emotional control was starting to fray with fatigue.

"We are concerned that the technique not be misused, so we are very keen to see legislation," he said. "We think it is important to effectively prohibit cloning in humans."

His temper flared. "We always talk [about the misuse of cloning] as if this will happen in some other country, never ours. I am quite surprised that we are so arrogant about our own level of morality and so disparaging about others.

"Little cloned copies of despots are not frightening. It is the big despot alive now that is frightening," he said. "Saddam Hussein is more frightening than any clone he could make."

The cloning experiment has highlighted the tensions between science and the society it seeks to serve, he acknowledged.

In his view, science is meant to exploit the new opportunities it creates, and its findings should be published as openly as possible. Secrecy is an anathema, as is any effort to stifle basic research, he said.

"If academic research does produce things that are useful, like therapeutic proteins or methods of embryo freezing, they should be taken and used," he said.

"The objectives we always had with this project were to produce methods for making precise genetic change in livestock and for making copies of embryos. It is legitimate to do things in farm animals that you would not do in humans.

"There is some suggestion that this work was not properly assessed before it was started. . . . How do you regulate things beforehand?" he said. "That is the antithesis of science. The only appropriate thing is to regulate something carefully after it has been published and described."

By the time Dolly was born, her progenitor—the language has yet to coin a more precise word for the creature that contributed her cell—was long dead. Her udder cells had been frozen and stored until they were used to produce the cloned sheep.

The researchers inserted the cells into 277 eggs to make clone embryos. Many survived the transfer into the wombs of other sheep, but only one survived to birth.

At tea time on July 5, 1996, Dolly came head-first into the world.

"I can't remember anything about the day the lamb was born," Wilmut said.

But no matter how many times he answers, the questions remain.

And there is no shortage of people waiting to ask them.

The phone is ringing again. The institute has answered 2,000 calls about cloning in less than two weeks. The fax spits out another letter of inquiry. The security gate is buzzing as a reporter seeks admission after closing hours.

On this night, Wilmut heads for the pub, where the man who has developed biotechnology's newest innovation indulges himself in one of its oldest—a pint of brown ale.

"Ian's about to enter meltdown," said Harry Griffin.

RICHARD SEED: MEDIA ATTENTION FEEDS THE CLAIMS OF A PUBLICITY SEEKER

Dirk Johnson

Physicist Richard Seed made headlines in December 1997 when he announced at a genetics conference that he would soon start cloning humans, *New York Times* writer Dirk Johnson explains in the following article. Johnson quotes ethicists and journalists who believe that subsequent media accounts, especially a National Public Radio report by Joe Palca, gave Seed's claim more publicity than it deserved and created undue public anxiety over human cloning. Most scientists feel that Seed is very unlikely to succeed in his aims, the author states, yet many of the media accounts treated Seed's claims seriously. Johnson also reports that members of Seed's family are unhappy with the coverage, which they believe exaggerated Seed's eccentricity and past career failures, portraying him almost as a madman.

As scientists and academics batted around some esoteric notions at a conference on genetics in December 1997, a 69-year-old bearded man stood up, made some vague and rambling statements, then issued a grandiose claim:

He would soon start cloning humans.

The Media Make a Dr. Frankenstein

In the next weeks that man, Dr. Richard Seed, would burst onto front pages across America, become the focus of network television shows, the lightning rod for angry speeches in Congress and stark warnings from the White House.

He became a modern-day Dr. Frankenstein whose plans helped frighten 19 European nations into signing a ban against duplicate baby making.

But more interesting than Seed's proposal is how an eccentric physicist with no job, no money and—most scientists agree—no chance of pulling off the sort of science-fictionlike cloning factory he

Excerpted from "Eccentric's Hubris Sets Off Global . . ." by Dirk Johnson, January 26, 1998. Copyright ©1998 by The New York Times. Reprinted with permission from *The New York Times*.

proposed became such a huge national figure.

Arthur Caplan, director of the Center for Bioethics at the University of Pennsylvania, who was the keynote speaker at the December conference on the ethical issues in human genetics where Seed made his startling proposal, said the real Dr. Frankenstein in this case was the news media.

Besides the shortcomings of news organizations in deciphering scientific assertions, experts say, the sudden stardom of Seed also speaks volumes about the nation's fascination and anxiety over genetic technology at a time when virtually everything seems technologically possible.

"One of the great subjects for journalistic review," Caplan said, "will be how this man with no money, no standing with physicists, no organizational skills—an oddball, really—how this man suddenly turns into this authority chatting on the nightly news.

"Seed was legitimated by the very people who should have been scrutinizing him."

Seed's Background

For his part, Seed seemed a recruit from central casting: an eccentric, older scientist with a beard, mischievous eyes and a penchant for making outrageous statements.

Even his name gave his procreative adventure a sense of destiny.

He had three degrees from Harvard—a bachelor's, a master's and a doctorate in physics.

Seed developed a technique in the 1980s to transfer fertilized embryos from one woman to another for a company he helped found, Fertility and Genetics Research Inc. in Chicago.

But technology soon leaped ahead and Seed's business faded, said Walter G. Cornett III, who served on the company's board of directors. After that, Seed went into mortgage financing but eventually lost money in the business.

Seed has tried a variety of get-rich ventures. He once came up with a scheme to corner the world's fish-meal market. He went to see Cornett, a venture capitalist in the North Shore suburbs of Chicago, and said he was looking for investments to acquire seven small fishing fleets.

"He plopped down in a chair in front of me," Cornett recalled, "and said, 'I'm the world's smartest man.'"

Cornett chuckled in response, he said, before realizing Seed was not joking.

"You could tell he was absolutely dead serious," Cornett said.

Seed said he was looking for an investment of $35 million. Cornett told him that whatever the merits of the idea—"and for all I know, he was right," the venture capitalist said—he was not going to be putting up $35 million.

Three years ago, Seed began research on whether blood transfusions could reduce rejections of transplanted organs. He rented office space at the University of Illinois at Chicago. Ultimately, the research proved fruitless.

The Start of a Media Frenzy

Perhaps the first account of Seed's plans to start a human cloning clinic appeared December 11, 1997, in *The Washington Times*, just days after his announcement at the scientific meeting in Chicago.

In a telephone interview with the newspaper, Seed was quoted as saying that he needed $2 million to begin his project, and that he thought that pregnancy through cloning "could be achievable in less than a year."

Seed leapt onto the national scene January 7, 1998, in a report by National Public Radio. Seed was interviewed extensively by NPR's science reporter, Joe Palca.

In the report, Palca noted the considerable hurdles that Seed would need to clear before opening a cloning clinic.

"Right now, he doesn't have the money; he doesn't have a firm commitment from the physicians who must perform the procedure," Palca noted, "and he doesn't have an infertile couple willing to undergo the procedure."

On the day the interview was broadcast, NPR frequently referred to Seed's cloning plan.

It was the imprimatur of the prestigious National Public Radio that gave the story of Seed a real boost. Within days, it was nearly impossible to pick up a newspaper or watch television news without hearing about the man with the plan to clone humans.

Palca, interviewed after the uproar that followed his report, defended the focus on Seed because, after all, he said, "If somebody was going to try this, it was going to be someone like Seed, not someone from the mainstream, establishment scientific community."

He said he felt his report was properly skeptical.

"I didn't say he was going to succeed," Palca said. "I said he was going to try. I decided that it was worth a story for NPR, and I did it."

The reporter added that he "wouldn't have pitched it the way" many other news organizations did later, which conveyed the impression that Seed was on the brink of cloning.

But others said Palca's report put in play a story that soon became a case of journalistic hysteria.

"American journalists helped Richard Seed scare the pants off everybody," wrote John Kass, a columnist at the *Chicago Tribune*. "Alleged responsible newsies hyped his claims as if they were carnival barkers at the freak show, selling tickets to the tent of Jo-Jo the Dog Faced Boy."

To be sure, many of the TV and newspaper accounts touched on Seed's long odds.

In a serious look into Seed's checkered past and scientific limitations, the *Chicago Tribune* noted that "what some had feared was a real-life science-fiction horror story was looking more like a sad comedy."

Madman or Victim?

The son of a prominent Chicago surgeon, Seed recently lost his home to foreclosure, and now lives in a modest Chicago house owned by two of his children, said a brother, Dr. Randolph Seed.

Randolph Seed has been a partner in some of his brother's ventures. Richard Seed is being supported by his wife, who is a secretary, said Randolph, a surgeon.

For now, Seed has stopped talking to interviewers about his cloning plans.

"He's exhausted and frazzled," said Randolph Seed, who thinks that his brother could be successful at cloning, but that he had been exploited by the news media and portrayed as a madman.

"Richard has never been very good at dealing with people," his brother said. "He just says what he thinks. So he was perfect for the media."

Randolph Seed said that family members, worried about how his brother was being portrayed in news accounts, formed "a little support group" that counseled Richard Seed and ultimately persuaded him to stay out of the limelight.

"They would chew him up and spit him out," Randolph Seed said of interviewers. But he insisted that if his brother found the financial backing, he could make good on the cloning plans.

And he said there are many true-believers and fans of this scientific quest.

"We've had calls from people who want to be cloned," Randolph Seed said.

CHAPTER 4

HUMAN CLONING
AND THE LAW

FEW LEGAL BARRIERS PREVENT HUMAN CLONING

Margaret A. Jacobs

The February 1997 announcement from Scotland that a lamb had been cloned from an adult ewe's mammary cell immediately raised questions about the ethics and legality of cloning humans. Writing a few days after the announcement, *Wall Street Journal* reporter Margaret A. Jacobs notes that there are few legal obstacles to such activities in the United States. However, she points out, in light of recent developments, some groups are likely to try to change that situation by promoting laws that would restrict or ban experiments in human cloning. Others, including many scientific researchers, would prefer the government to establish guidelines and regulations concerning human cloning rather than legislation, Jacobs reports. Since the time that this article was published, numerous bills banning human cloning have been introduced into Congress, but none has passed. However, California passed a law against human cloning in January 1998, and other states may do so as well.

There are few legal obstacles to the cloning of animals or even humans in the U.S., but many ethical questions and commercial disputes are likely to arise if companies start to apply the technology.

Bioethicists say groups representing doctors, farmers and churches are likely to object to any attempts to clone human beings and possibly any type of mammal. They expect them to lobby Congress to pass legislation to prohibit or strictly regulate the process. Indeed, the Foundation on Economic Trends said on February 24, 1997, that it had organized 400 religious and health organizations world-wide to push for new laws banning human cloning.

In March 1997, President Bill Clinton responded to news that researchers in Scotland had successfully cloned an adult sheep by asking the National Bioethics Advisory Commission, a group he created in 1995, to review the implications. While the discovery offers "potential benefits" for animal and drug research, the president said in a letter to the commission's chairman that it raises "serious ethical" questions if the technology is used on humans. He asked the commission to offer recommendations within 90 days on possible federal action to curb abuse.

"When all is said and done, it's an issue of ethics, not patent law: What sort of society do we want to have?" says Prof. Roger E. Schecter, an intellectual property professor at the George Washington University Law School in Washington, D.C.

Several roadblocks to commercial development of new medical and biological technologies have already been thrown up around the world. The European Community currently prohibits anyone from patenting genetically altered animals. And in 1996, the American Medical Association successfully pushed Congress to pass legislation watering down the rights of holders of patents to medical procedures. Federal law currently prohibits federal funds from being used to pay for human embryo research, including cloning.

Because of the funding ban, some bioethicists worry that government oversight of any human cloning research will be inadequate without further federal or state action. Alexander Morgan Capron, a member of the National Bioethics Advisory Commission, frets that fertility centers might begin experimenting with cloning if desperate patients are willing to pay for the research.

Some lawyers who specialize in biotechnology say the government should regulate cloning gingerly, if it does so at all. "We would rather have the National Institutes of Health or other scientific group issue guidelines than have Congress step in and regulate," says Teresa Stanek Rea, an Alexandria, Va., patent lawyer and member of the board of the American Intellectual Property Association, a trade group.

Patents and Clones

Still, lawyers say researchers currently should be able to get a patent on the cloning process in the U.S., which would give the holder an effective monopoly on the technique for 20 years from the date an application is filed.

In a 1980 decision, the U.S. Supreme Court said the Constitution doesn't prohibit the patenting of living creatures. The court upheld a patent on a single-celled bacterium developed to clean up oil spills. Since that decision, the U.S. Patent and Trademark Office has granted patents on the genes used to produce 29 different types of genetically altered animals, including 23 mice, one rabbit, one sheep, one rat, one bird, one fish and a worm. The animals are used primarily in research, according to Lisa-Joy Zgorski, a spokeswoman for the Patent Office.

The patent provides ownership of the roadmap to understand the gene rather than the gene itself. "We're not in the business of moral judgment," says Ms. Zgorski.

A Ban on U.S. Funds for Human Cloning Research

Marlene Cimons and Jonathan Peterson

On March 4, 1997, about two weeks after the announcement that a lamb had been cloned from an adult sheep cell, U.S. president Bill Clinton announced a ban on federal funding for research on human cloning, pending the findings of the National Bioethics Advisory Commission. Clinton also asked privately funded scientists not to carry on such research. *Los Angeles Times* staff writers Marlene Cimons and Jonathan Peterson recount Clinton's decision and the reaction to it. They point out that, in fact, no federal funds were being spent on human cloning projects at the time of Clinton's announcement; indeed, no federal funding could be spent on any kind of research involving human embryos. Clinton's action thus seemed designed to calm public fears of a hasty entry into research concerning human cloning, they note. The authors also write that many ethicists and scientists approved the temporary ban, with hopes that it would provide a period of time for society to fully consider and debate the implications of human cloning.

Stepping into an uncharted intersection of science and morality, President Bill Clinton on March 4, 1997, banned the use of federal funds for human cloning research and called upon private sector scientists to voluntarily refrain from such experiments.

Responding to a February report that a Scottish scientist had cloned a sheep using genetic material from an adult sheep—and slightly later news of the cloning of two monkeys in Oregon—Clinton cautioned that the emerging science is creating new ethical burdens for humanity even as it holds great promise for agriculture, medicine and other areas of commerce.

A Need for Caution

"Science often moves faster than our ability to understand its implications," said Clinton, who chose to ban funding for human cloning work while a special presidential bioethics panel studies the issues.

"That is why we have a responsibility to move with caution and care" to harness the emerging technology, he said.

"There is much about cloning that we still do not know," he added. "But this much we do know: Any discovery that touches upon human creation is not simply a matter of scientific inquiry, it is a matter of morality and spirituality as well."

Members of the National Bioethics Advisory Commission were expected to report back to the president in the spring of 1997.

Clinton's action appeared to have more psychological impact than immediate scientific significance.

The National Institutes of Health, which provides the bulk of research money to U.S. scientists, does not now support any research projects involving human cloning.

Furthermore, as part of the 1996–97 legislation reauthorizing NIH, Congress explicitly prohibited any federally funded human embryo research.

Also, in 1994, Clinton banned the use of federal money to support the creation of human embryos solely for research purposes.

Clinton said the purpose of his action was to close any possible loopholes in existing policy that still might allow research on human cloning to go forward.

The order does not affect animal cloning research.

"My own view is that human cloning would have to raise deep concerns, given our most cherished concepts of faith and humanity," Clinton said.

"Each human life is unique, born of a miracle that reaches beyond laboratory science," Clinton said as he issued the executive directive.

"I believe we must respect this profound gift and resist the temptation to replicate ourselves," he said.

"At the very least, however, we should all agree that we need a better understanding of the scope and implications of this most recent breakthrough."

No Longer Science Fiction

Long grist to science fiction's mill and a distinctly distant future, the idea of cloning human beings abruptly seemed more plausible in February 1997 when Scottish scientist Ian Wilmut announced that he had succeeded in cloning a lamb named Dolly, which since has grown into a healthy adult—a genetic carbon copy of the single adult sheep that provided the genetic material.

Several days later, it was revealed that scientists in Oregon had cloned from embryonic cells two rhesus monkeys, a species much closer to that of humans.

Cloning is the production of an exact genetic duplicate of a living organism.

In normal sexual reproduction, an egg and a sperm—each containing half the genetic complement of an adult—fuse, combining their DNA to produce the complete genetic blueprint of a third adult.

In cloning, however, all of the genetic material comes from one parent, and the offspring is genetically identical to that parent.

NIH Director Harold E. Varmus said that the president "was trying to provide some reassurance to the public that federal monies are not being used to do specific cloning of human beings," thus allowing the commission "time to think things through."

"This should calm people's fears about those nightmarish possibilities that are extremely unlikely, and get them to focus on the real dilemmas," Varmus added.

Art Caplan, director of the center for bioethics at the University of Pennsylvania, agreed, calling Clinton's move a sensible approach to a volatile scientific issue.

Call for a Pause

Human cloning research—at this time—"is too risky, too dangerous to undertake," Caplan said. "We're only at the Wright Brothers stage of development with respect to cloning technology," he said.

"A sheep is in a barn, and a monkey is in a cage," he said. But to reach that point, "a number of dead embryos and deformed animals were made as well. This is not a technique that is ready right now for human application. It makes sense to impose a moratorium and let society catch its collective moral breath."

Clinton's move did not seem to provoke the usual tension that results when a politician intervenes in scientific matters—further indication, perhaps, of the widespread recognition that cloning research is a moral minefield.

Political involvement in scientific research is neither new nor has it been partisan in nature.

Clinton's two predecessors, George Bush and Ronald Reagan, both Republicans, maintained a ban sought by many in the anti-abortion movement on federally funded research using fetal tissue.

Rep. John D. Dingell (D-Mich.) rattled the biomedical community by initiating a series of scientific fraud and abuse cases when he chaired the House Energy and Commerce subcommittee on oversight and investigations.

Over the past two years, the friction between science and politics accelerated further under the Republican Congress, with GOP lawmakers taking legislative aim at certain research areas, such as work involving human embryos, that they didn't like.

Varmus said he believed Clinton's action would have the opposite effect.

"It takes the pressure off any need to legislate," he said.

Bioethicist Caplan agreed. "Some may cry, 'Censorship,' but it's simply silly to think that research this controversial, and that has such potential for misuse, is not going to elicit a political response."

In addition to a ban on federal funding, Clinton called upon the private sector to voluntarily refrain from research in this area until the national debate is concluded.

Driving Science Underground?

It was unclear how privately funded scientists would respond.

Varmus acknowledged that scientists in the private sector "probably will not be happy about it," but predicted most would respect it. "I can't think that there is much commercial interest in this subject."

Varmus added: "It's impossible to exclude the science-fiction rogue scientist idea. But it's very hard to do this stuff, and it's remote that anything could happen in 90 days. I don't think this is imminent."

Nevertheless, embryology experts speculated that one or more researchers at in-vitro fertilization clinics might already be engaged in some form of exploratory research, although none would comment publicly.

IVF clinics, which receive little, if any, federal funding, are supported by client fees and a limited amount of private funding.

Other experts predicted that a long-term ban could prompt some U.S. researchers to move that element of their research abroad.

When Australia and other countries banned IVF research, for example, scientists left for more amenable countries like Singapore.

But for all his caveats, Clinton agreed with numerous experts who have said that the Scottish sheep cloning held the potential for stunning benefits in medical applications, food production, and even the saving of endangered species.

"The recent breakthrough in animal cloning is one that could yield enormous benefits, enabling us to reproduce the most productive strains of crop and livestock, holding out the promise of revolutionary new medical treatments and cures, helping to unlock the greatest secrets of the genetic code," Clinton said.

"But like the splitting of the atom, this is a discovery that carries burdens as well as benefits."

THE NATIONAL BIOETHICS ADVISORY COMMISSION'S EVALUATION OF HUMAN CLONING

Harold T. Shapiro

In early March 1997, just days after scientists in Scotland announced that they had cloned a sheep from an adult cell and thus led some commentators to suggest that cloning of humans might be imminent, U.S. president Bill Clinton asked the National Bioethics Advisory Commission to report to him on the legal and ethical issues raised by human cloning. Harold T. Shapiro, the head of the commission, describes in the following selection the process that the commission used for its review and the conclusions it reached. The commission's primary conclusion, Shapiro relates, is that the cloning procedure cannot be considered safe for human fetuses at this time, and therefore a moratorium should be placed on research into human cloning. However, he recommends that any future laws about human cloning be drafted carefully so as not to block promising medical research related to cloning. Shapiro is an economist and the president of Princeton University in New Jersey.

The idea that humans might someday be cloned from a single adult somatic cell without sexual reproduction moved further away from science fiction and closer to a genuine possibility when scientists at the Roslin Institute in Scotland announced the successful cloning of a sheep by a new technique that had never before been fully successful in mammals. The technique involved transplanting the genetic material of an adult sheep, apparently obtained from a well-differentiated somatic cell, into an egg from which the nucleus had been removed. The resulting birth of the sheep, named Dolly, on 5 July 1996, was different from prior attempts to create identical offspring because Dolly contained the genetic material of only one parent and was therefore a "delayed" genetic twin of a single adult sheep.

This cloning technique, which I will refer to as "somatic cell

Excerpted from "Ethical and Policy Issues of Human Cloning," by Harold T. Shapiro, *Science*, July 11, 1997. Copyright ©1997 by the American Association for the Advancement of Science. Reprinted with permission from *Science*.

nuclear transfer," is an extension of research that had been going on for over 40 years with nuclei derived from nonhuman embryonic and fetal cells. The further demonstration that nuclei from cells derived from an adult animal could be "reprogrammed," or that the full genetic complement of such a cell could be reactivated well into the chronological life of the cell, is what sets the results of this experiment apart from prior work. At the same time, several serious scientific uncertainties remain that could have a significant impact on the potential ability of this new technique to create human beings. Examples of such uncertainties include the impact of genetic imprinting, the nature of currently unknown species differences, and the effects of cellular aging and mutations.

The initial public response to this news, here and abroad, was primarily one of concern. In some cases, these concerns were amplified by largely fictional and mistaken accounts of how this new technology might dramatically reshape the future of our society. The sources of these feelings were complex, but usually centered around the basic fact that this technique would permit human procreation in an asexual manner, would allow for an unlimited number of genetically identical offspring, and would give us the capacity for complete control over the genetic profile of our children.

Ethical and Legal Issues

Within days of the published report, President Bill Clinton instituted a ban on federal funding related to attempts to clone human beings in this manner. In addition, the president asked the recently appointed National Bioethics Advisory Commission (NBAC) to report within 90 days on the ethical and legal issues that surround the potential cloning of human beings.

This was an unusually challenging assignment for many reasons. These issues are complex and difficult, and many scientific uncertainties remain. Conflicting values are at stake, and Americans disagree on the implications of this new technology for the social and cultural values they hold dearest. It is difficult to decide if and when our liberties, including the freedom of scientific inquiry, should be restricted. Finally, the commission was given an ambitious timetable.

Nonetheless, NBAC made every effort to consult widely with ethicists, theologians, scientists, scientific societies, physicians, and others in initiating an analysis of the many scientific, legal, religious, ethical, and moral dimensions of the issue. This included a careful consideration of the potential risks and benefits of using this technique to create children and a review of the potential constitutional challenges that might be raised if new legislation were to restrict the creation of a child through somatic cell nuclear transfer cloning.

The commission focused its attention on the new and distinctive ethical issues that would be raised by the use of this technique for the purpose of creating an embryo genetically identical to an existing (or previously existing) person that would then be implanted in a woman's uterus and brought to term. Although the creation of embryos for research purposes alone always raises serious ethical questions, these issues have recently received extensive analysis and deliberation in our country, and the use of somatic cell nuclear transfer to create embryos raises no new issues in this respect. The unique and distinctive ethical issues raised by the use of somatic cell nuclear transfer to create children relate to serious safety concerns and to a set of questions about what it means to be human; questions that go to the heart of the way we think about families and relationships between generations, our concept of individuality, and the potential for treating children as objects, as well as issues of constitutional law that might be involved in the area of procreation.

Concerns

In its deliberations, NBAC reviewed the scientific developments that preceded the Roslin announcement, as well as those likely to follow in its path, and the many moral and legal concerns raised by the possibility that this technique could be used to clone human beings. Although some of the initial negative response arose from fictional accounts of cloning human beings, more thoughtful concerns revealed fears about harm to the children who may be created in this manner, particularly psychological harm associated with a possibly diminished sense of individuality and personal autonomy. Others expressed concern about a degradation in the quality of parenting and family life.

In addition to concerns about specific harms to children, people have frequently expressed fears that the widespread practice of somatic cell nuclear transfer cloning would undermine important social values by opening the door to a form of eugenics or by tempting some to manipulate others as if they were objects instead of persons. These are concerns worthy of widespread and intensive debate, but arrayed against these concerns are other vitally important social and constitutional values, such as protecting the widest possible sphere of personal choice, particularly in matters pertaining to procreation and child rearing; maintaining privacy; protecting the freedom of scientific inquiry; and encouraging the possible development of new biomedical breakthroughs.

To arrive at its recommendations, NBAC also examined longstanding religious traditions and found that religious positions on human cloning are pluralistic in their premises, modes of argument, and conclusions. Some religious thinkers argue that the use of somatic cell nuclear transfer cloning to create a child would be intrinsically immoral and thus could never be morally justified. Other religious thinkers contend that human cloning to create a child could be morally justified under some circumstances but believe that it should be strictly regulated to prevent abuses.

Public Policies

The public policies that NBAC recommended with respect to the creation of a child by means of somatic cell nuclear transfer reflected the commission's attempt to balance the various interests at stake and to apply its best judgments about the ethics of attempting such an experiment at this time as well as its view of U.S. constitutional traditions regarding limitations on individual actions in the name of the common good. We concluded that, at present, the use of this technique to create a child would be a premature experiment that would expose the fetus and the developing child to unacceptable risks. In our judgment, this in itself might be sufficient to justify a prohibition on using this new technique to clone human beings at this time, even if such efforts were to be characterized as the exercise of a fundamental right to attempt to procreate. Beyond the issue of the safety of the procedure, however, NBAC found that concerns relating to potential psychological harm to children and effects on the moral, religious, and cultural values of society merit further reflection and deliberation. Whether upon such further deliberation our nation will conclude that the use of this new cloning technique to create children should be allowed or permanently banned is, for the moment, an open question. Fortunately, time is an ally in this regard, allowing for the accrual of further data from animal experimentation, an assessment of the prospective safety and efficacy of the procedure in humans, and a period of fuller national debate on ethical and social concerns.

The commission therefore concluded that a period of time should be imposed in which no attempt is made to create a child using somatic cell nuclear transfer.

Recommendations Concerning Legislation

Within this overall framework, the commission's full set of conclusions and recommendations was as follows:

1) The commission concluded that at this time it is morally unacceptable for anyone in the public or private sector, whether in a research or clinical setting, to attempt to create a child using somatic cell nuclear transfer cloning. We reached a consensus on this point because current scientific information indicates that this technique is not safe to use in humans at this time. Indeed, we believe that it would violate important ethical obligations were clinicians or researchers to attempt to create a child using these particular technologies, which are likely to involve unacceptable risks to the fetus or potential child. Moreover, in addition to safety concerns, many other serious ethical concerns have been identified that require much more widespread and

careful public deliberation before this technology may be used.

The commission therefore recommended the following: (i) A continuation of the current moratorium on the use of federal funding to support any attempt to create a child by somatic cell nuclear transfer. (ii) An immediate request to all firms, clinicians, investigators, and professional societies in the private and nonfederally funded sectors to comply voluntarily with the intent of the federal moratorium. Professional and scientific societies should make clear that any attempt to create a child by somatic cell nuclear transfer and implantation into a woman's body would at this time be an irresponsible, unethical, and unprofessional act.

2) The commission further recommended that federal legislation should be enacted to prohibit anyone from attempting, whether in a research or clinical setting, to create a child through cloning by somatic cell nuclear transfer. It is critical, however, that such legislation include a sunset clause to ensure that Congress will review this issue after a specified period of time (3 to 5 years) to decide whether the prohibition continues to be needed. If state legislation is enacted, it should also contain such a sunset provision. Any such legislation or associated regulation should require that at some point before the expiration of the sunset period, an appropriate oversight body will evaluate and report on the current status of somatic cell nuclear transfer technology and on the ethical and social issues that its potential use to create human beings would raise in the light of public understandings at that time.

3) The commission also concluded that (i) any regulatory or legislative actions undertaken to effect the foregoing prohibition should be carefully written so as not to interfere with other important areas of scientific research. In particular, we believe that no new regulations are required regarding the cloning of human DNA sequences and cell lines, because neither activity raises the scientific and ethical issues that arise from the attempt to create children through somatic cell nuclear transfer, and these fields of research have already provided important scientific and biomedical advances. Likewise, research on cloning animals by this technique does not raise the same issues as attempting to use it for human cloning, and its continuation should only be subject to existing regulations regarding the humane use of animals and to review by institution-based animal protection committees. (ii) If a legislative ban is not enacted, or is enacted but later lifted, clinical use of somatic cell nuclear transfer techniques to create a child should be preceded by research trials that are governed by the twin protections of independent review and informed consent, which is consistent with existing norms of human subjects protection. (iii) The U.S. government should cooperate with other nations and international organizations to enforce any common aspects of their respective policies on the cloning of human beings.

Discussion and Education

4) The commission concluded that different ethical and religious perspectives and traditions are divided on many of the important moral issues that surround this topic. Therefore, it recommended that the federal government and all interested and concerned parties encourage widespread and continuing deliberation on these issues to further our understanding of the ethical and social implications of this technology and to enable society to produce appropriate long-term policies should the time come when present concerns about safety have been addressed.

5) Finally, because scientific knowledge is essential for all citizens to participate in a full and informed fashion in the governance of our complex society, the commission recommended that federal departments and agencies concerned with science should cooperate in seeking out and supporting opportunities to provide information and education to the public in the area of genetics and about other developments in the biomedical sciences, especially where they affect important cultural practices, values, and beliefs.

NBAC hopes that the sections of its report that outline the scientific, religious, ethical, and legal issues associated with human cloning will form a useful basis for the widespread deliberations and broad public education we believe are so essential. We believe that this kind of deliberation and education are especially critical in a society where individuals hold various religious and moral perspectives. As I have already noted, issues related to human cloning in this novel manner go to the very nature of what it means to be human and to the very heart of what people think of as their families and their individuality. These are issues worthy of intensive and widespread debate.

Once again, however, time is an ally allowing for the accumulation of more scientific data from animal studies as well as granting an opportunity for fuller national debate on ethical and moral concerns. Through such deliberation, we can, as a society, improve not only our understanding of the scientific issues but our prospects for achieving moral agreement where that is possible, or mutual respect where such agreement cannot be achieved.

IS A BAN ON HUMAN CLONING CONSTITUTIONAL?

Mark D. Eibert

Mark D. Eibert is an attorney who practices in San Mateo, California. In the following selection, Eibert examines what he sees as a disturbing trend favoring strict laws against human cloning and all research related to human cloning. Eibert notes that California banned human cloning in 1998, and several other states as well as the U.S. Congress are considering similar laws. According to Eibert, such bans on human cloning would be unconstitutional. The Supreme Court has ruled that all Americans have a constitutional right to "bear or beget" children, including by artificial means such as in vitro fertilization, he points out. This precedent should also apply to the creation of children through cloning technology, he asserts, which would be a boon to thousands of infertile couples. Furthermore, Eibert suggests that the constitutional right of free speech may be interpreted as including scientific inquiry, which would mean that legislation banning human cloning research would violate this right.

"There are some avenues that should be off limits to science. If scientists will not draw the line for themselves, it is up to the elected representatives of the people to draw it for them."

Thus declared Sen. Christopher "Kit" Bond (R-Mo.) one of the sponsors of S. 1601, the official Republican bill to outlaw human cloning. The bill would impose a 10-year prison sentence on anyone who uses "human somatic cell nuclear transfer technology" to produce an embryo, even if only to study cloning in the laboratory. If enacted into law, the bill would effectively ban all research into the potential benefits of human cloning. Scientists who use the technology for any reason—and infertile women who use it to have children—would go to jail.

Not to be outdone, Democrats have come up with a competing bill. Sens. Ted Kennedy (Mass.) and Dianne Feinstein (Calif.) have proposed S. 1602, which would ban human cloning for at least 10 years. It would allow scientists to conduct limited experiments with cloning in the laboratory, provided any human embryos are destroyed at an early stage rather than implanted into a woman's uterus and allowed to be born.

If the experiment goes too far, the Kennedy-Feinstein bill would impose a $1 million fine and government confiscation of all property, real or personal, used in or derived from the experiment. The same penalties that apply to scientists appear to apply to new parents who might use the technology to have babies.

FDA and State Cloning Bans

The near unanimity on Capitol Hill about the need to ban human cloning makes it likely that some sort of bill will be voted on the 1998–1999 congressional session and that it will seriously restrict scientists' ability to study human cloning. In the meantime, federal bureaucrats have leapt into the breach. In January 1998, the U.S. Food and Drug Administration (FDA) announced that it planned to "regulate" (that is, prohibit) human cloning. In the past, the FDA has largely ignored the fertility industry, making no effort to regulate in vitro fertilization, methods for injecting sperm into eggs, and other advanced reproductive technologies that have much in common with cloning techniques.

An FDA spokesperson told me that although Congress never expressly granted the agency jurisdiction over cloning, the FDA can regulate it under its statutory authority over biological products (like vaccines or blood used in transfusions) and drugs. But even Rep. Vernon Ehlers (R-Mich.), one of the most outspoken congressional opponents of cloning, admits that "it's hard to argue that a cloning procedure is a drug." Of course, even if Congress had granted the FDA explicit authority to regulate cloning, such authority would only be valid if Congress had the constitutional power to regulate reproduction—which is itself a highly questionable assumption (more on that later).

Nor have state legislatures been standing still. Effective January 1, 1998, California became the first state to outlaw human cloning. California's law defines "cloning" so broadly and inaccurately—as creating children by the transfer of nuclei from any type of cell to enucleated eggs—that it also bans a promising new infertility treatment that has nothing to do with cloning. In that new procedure, doctors transfer nuclei from older, dysfunctional eggs (not differentiated adult cells as in cloning) to young, healthy donor eggs, and then inseminate the eggs with the husband's sperm—thus conceiving an ordinary child bearing the genes of both parents. Taking California as their bellwether, many other states are poised to follow in passing very restrictive measures.

Does Congress Have the Right to Ban Cloning?

What started this unprecedented governmental grab for power over both human reproduction and scientific inquiry? Within days after Dolly, the cloned sheep, made her debut in February 1997, President Bill Clinton publicly condemned human cloning. He opined that "any discovery that touches upon human creation is not simply a matter of scientific inquiry. It is a matter of morality and spirituality as well. Each human life is unique, born of a miracle that reaches beyond laboratory science."

Clinton then ordered his National Bioethics Advisory Commission to spend all of 90 days studying the issue—after which the board announced that it agreed with Clinton. Thus, Clinton succeeded in framing the debate this way: Human cloning was inherently bad, and the federal government had the power to outlaw it.

But in fact, it's far from clear that the government has such far-reaching authority. Several fundamental constitutional principles conflict with any cloning ban. Chief among them are the right of adults to have children and the right of scientists to investigate nature.

The Supreme Court has ruled that every American has a constitutional right to "bear or beget" children. This includes the right of infertile people to use sophisticated medical technologies like in vitro fertilization. As the U.S. District Court for the Northern District of Illinois explained, "Within the cluster of constitutionally protected choices that includes the right to have access to contraceptives, there must be included . . . the right to submit to a medical procedure that may bring about, rather than prevent, pregnancy."

About 15 percent of Americans are infertile, and doctors often cannot help them. Federal statistics show that in vitro fertilization and related technologies have an average national success rate of less than 20 percent. Similarly, a *Consumer Reports* study concludes that fertility clinics produce babies for only 25 percent of patients. That leaves millions of people who still cannot have children, often because they can't produce viable eggs or sperm, even with fertility drugs. Until recently, their only options have been to adopt or to use eggs or sperm donated by strangers.

Once cloning technology is perfected, however, infertile individuals will no longer need viable eggs or sperm to conceive their own genetic children—any body cell will do. Thus, cloning may soon offer many Americans the only way possible to exercise their constitutional right to reproduce. For them, cloning bans are the practical equivalent of forced sterilization.

In 1942, the Supreme Court struck down a law requiring the sterilization of convicted criminals, holding that procreation is "one of the basic civil rights of man," and that denying convicts the right to have children constitutes "irreparable injury" and "forever deprived [them] of a basic liberty." To uphold a cloning ban, then, a court would have to rule that naturally infertile citizens have less right to try to have children than convicted rapists and child molesters do.

What Harm Will Human Cloning Do?

Many politicians and bureaucrats who want to ban human cloning say they need their new powers to "protect" children. Reciting a long list of speculative harms, ranging from possible physical deformities to the psychic pain of being an identical twin, they argue in essence that cloned children would be better off never being born at all.

But politicians and the media have grossly overstated the physical dangers of cloning. The Dolly experiment started with 277 fused eggs, of which only 29 became embryos. All the embryos were transferred to 13 sheep. One became pregnant, with Dolly. The success rate per uterine transfer (one perfect offspring from 13 sheep, with no miscarriages) was better than the early success rates for in vitro fertilization. Subsequent animal experiments in Wisconsin have already made the process much more efficient, and improvements will presumably continue as long as further research is allowed.

More fundamentally, the government does not have the constitutional authority to decide who gets born—although it once thought it did, a period that constitutes a dark chapter of our national heritage. In the early 20th century, 30 states adopted eugenics laws, which required citizens with conditions thought to be inheritable (insanity, criminal tendencies, retardation, epilepsy, etc.) to be sterilized—partly as a means of "protecting" the unfortunate children from being born.

In 1927, the U.S. Supreme Court upheld such a law, with Oliver Wendell Holmes writing for the majority, "It would be better for all the world, if instead of waiting to execute degenerate offspring for crime, or to let them starve for their imbecility, society can prevent those who are manifestly unfit from continuing their kind. . . . Three generations of imbeciles are enough."

A Return to Eugenics?

California's eugenics law in particular was admired and emulated in other countries—including Germany in 1933. But during and after World War II, when Americans learned how the Nazis had used their power to decide who was "perfect" enough to be born, public and judicial opinion about eugenics began to shift. By the 1960s, most of the eugenics laws in this country had been either repealed, fallen into disuse, or were struck down as violating constitutional guarantees of due process and equal protection.

Indeed, those old eugenics laws were a brief deviation from an American tradition that has otherwise been unbroken for over 200 years. In America, it has always been the prospective parents, never the government, who decided how much risk was acceptable for a mother and her baby—even where the potential harm was much more certain and serious than anything threatened by cloning.

Hence, in vitro fertilization and fertility drugs are legal, even though they create much higher risks of miscarriages, multiple births, and associated birth defects. Individuals who themselves have or are known carriers of serious inheritable mental or physical defects such as sickle cell anemia, hemophilia, cystic fibrosis, muscular dystrophy, and Tay-Sachs disease are allowed to reproduce, naturally and through in vitro fertilization, even though they risk having babies with serious, or even fatal, defects or diseases. Older mothers at risk of having babies with Down Syndrome, and even women with AIDS, are also allowed to reproduce, both naturally and artificially. Even if prenatal testing shows a fetus to have a serious defect like Down Syndrome, no law requires the parents to abort it to save it from a life of suffering.

In short, until science revealed that human cloning was possible, society assumed that prospective parents could decide for themselves and their unborn children how much risk and suffering were an acceptable part of life. But in the brave new world of the federal bureaucrat, that assumption no longer holds true.

Ironically, some cloning opponents have turned the eugenics argument on its head, contending that cloning could lead to "designer children" and superior beings who might one day rule mankind. But allowing infertile individuals to conceive children whose genome is nearly identical to their already existing genomes no more creates "designer children" than it creates "designer parents." More important, these opponents miss the point that only government has the broad coercive power over society as a whole necessary to make eugenics laws aimed at "improving the race." It is those who support laws to ban cloning who are in effect urging the passage of a new eugenics law, not those who want to keep government out of the business of deciding who is perfect enough or socially desirable enough to be born.

Religious Arguments

Another significant driving force behind attempts to restrict or reverse an expansion in human knowledge stems from religious convictions. Interestingly, there is no necessary theological opposition to cloning: For example, two leading rabbis and a Muslim scholar who testified before the National Bioethics Advisory Commission had no objection to the practice and even advanced religious arguments for cloning.

Still, politicians from both major parties have already advanced religious arguments against cloning. President Clinton wants to outlaw cloning as a challenge to "our cherished concepts of faith and humanity." House Majority Leader Dick Armey also opposes cloning, saying that "to be human is to be made in the image and likeness of a loving God," and that "creating multiple copies of God's unique handiwork" is bad for a variety of reasons. Senator Bond warns that "humans are not God and they should not be allowed to play God"—a formulation similar to that of Albert Moraczewski, a theologian with the National Conference of Catholic Bishops, who told the president's commission, "Cloning exceeds the limits of the delegated dominions given to the human race."

Of course, virtually every major medical, scientific, and technological advance in modern history was initially criticized as "playing God." To give just two recent examples, heart transplants and "test-tube babies" both faced religious opposition when first introduced. Today, heart transplants save 2,000 lives every year, and in vitro fertilization helped infertile Americans have 11,000 babies in 1995 alone.

Religious belief doesn't require opposition to these sorts of expansions in human knowledge and technology. And basing a cloning ban primarily on religious grounds would seem to violate the Establishment Clause [of the First Amendment to the Constitution, which forbids Congress to make laws concerning religion]. But that's not the only potential constitutional problem with a ban.

A Right to Scientific Inquiry

Many courts and commentators say that a constitutional right of scientific inquiry is inherent in the rights of free speech and personal liberty. To be sure, certain governmental attempts to restrict the methods scientists can use have been upheld—for example, regulations requiring free and informed consent by experimental subjects. But those have to do with protecting the rights of others. Cloning bans try to stop research that everyone directly concerned wants to continue. As one member of the National Bioethics Advisory Commission observed, if the group's recommendation to ban cloning is enacted, it would apparently be the first time in American history that an entire field of medical research has been outlawed.

Prohibiting scientific and medical activities would also raise troubling enforcement issues. How exactly would the FBI—in its new role as "reproductive police" and scientific overseer—learn, then prove, that scientists, physicians, or parents were violating the ban? Would they raid research laboratories and universities? Seize and read the private medical records of infertility patients? Burst into operating rooms with their guns drawn? Grill new mothers about how their babies were conceived? Offer doctors reduced sentences for testifying against the patients whose babies they delivered? And would the government really confiscate, say, Stanford University Medical Center, if one of its many researchers or clinicians "goes too far"?

Fantasies Bear Bitter Fruit

The year since the announcement of Dolly's birth has seen unprecedented efforts by government to expand its power over both human reproduction and science. Decisions traditionally made by individuals—such as whether and how to have children, or to study the secrets of nature—have suddenly been recast as political decisions to be made in Washington.

Human cloning, when it actually arrives, is not apt to have dire consequences. Children conceived through cloning technology will be not "Xerox copies" but unique individuals with their own personalities and full human rights. Once this basic fact is understood, the only people likely to be interested in creating children through cloning technology are incurably infertile individuals. There are already tens of millions of identical twins walking the earth, and they have posed no threat so far to God, the family, or country. A few more twins, born to parents who desperately want to have, raise, and love them as their own children, will hardly be noticed.

As for the nightmare fantasies spun by cloning opponents, even the president's special commission has admitted that fears of cloning being used to create hordes of Hitlers or armies of identical slaves are "based . . . on gross misunderstandings of human biology and psychology." And laws already prohibit criminal masterminds from holding slaves, abusing children, or cutting up people for spare body parts.

As harmless as the fact of cloning may be, the fear of cloning is already bearing bitter fruit: unprecedented extensions of government power, based either on unlikely nightmare scenarios or on an unreasoning fear that humans were "not meant" to know or do certain things. Far from protecting the "sanctity" of human life, such an attitude, if consistently applied, would doom the human race to a "natural" state of misery.

REGULATING HUMAN CLONING

Gregory E. Pence

Gregory E. Pence is a professor of philosophy at the University of Alabama, Birmingham. The following selection is taken from his book *Who's Afraid of Human Cloning?* Regulation is a more effective answer to the ethical dilemmas presented by the cloning of humans than a ban would be, Pence concludes. He suggests that the U.S. Centers for Disease Control draw up guidelines that would regulate fertility clinics using human cloning procedures. Such regulations could be modeled on those that have been used successfully to govern genetic engineering and human gene therapy, Pence says. He recommends that these regulations be reviewed and revised periodically as research into human cloning advances.

I believe it would be a mistake to pass laws banning nuclear somatic transfer (NST). Because these issues are linked in the minds of some to abortion and to attacks on motherhood, they inflame too many passions. As such, once human reproduction by NST is banned, it will be very difficult to later undo such a ban.

The National Bioethics Advisory Committee (NBAC) wants such legislation, but with a sunset clause. The history of past attempts to get such a qualification is not impressive. The 1994 Human Embryo Panel wanted Congress to accept some embryo research in carefully-defined areas. What it got was a blanket rejection of all such research. Some issues defy legislation with careful distinctions, and originating humans by NST is one such area.

On the other hand, I think regulation done in a limited way might work and would be justified. The best way would not be a new federal agency but bringing infertility clinics under the loose supervision of the Centers for Disease Control. After some conferences to get input from all sides, guidelines could be made for such clinics and revised every three years.

Model Regulation on Gene Therapy

Robert Cook Deegan, Director of the National Policy Board for the Institute of Medicine of the National Academy of Science, argues that we should model the control of human asexual reproduction on the model of gene therapy, which has successfully supervised genetic therapy over the last decade by regulation without legislation. I think he is correct. Initially feared, somatic genetic therapy is on its way to becoming an accepted part of medicine.

Through regulation of in vitro fertilization (IVF) clinics, two important questions can be discussed and answered: (1) When is it safe to originate a child by NST? and (2) Will it work when we try to do so? Everyone can have their say about this in public discussions, including scientists and representatives of religions.

There is a great historical model for how such regulation can be successful. The regulation of recombinant DNA research, the big controversy of the late 1970s, was contained by voluntary efforts of leading scientists in convening the famous Asimolar (California) conference in February 1975. Such research now is only supervised by local Institutional Review Boards (IRBs) and the National Institutes of Health (NIH) Recombinant DNA Advisory Committee (RAC), which advises NIH on the regulation of gene therapy. The RAC has functioned as a national ethics committee for a particular modality of treatment, viz., for human somatic gene therapy. Under this plan, no abuses have occurred and the public opinion now accepts somatic gene therapy.

To be fair, there are those who see things differently. Some researchers believe that RAC review has been way too cautious and too slow. People now see somatic gene therapy as just another kind of medicine. Hundreds of cancer protocols go forward with much less review. Isn't it time, critics ask, to loosen up RAC review of somatic gene therapy? Especially where the Federal Drug Administration must also review their protocols, subjecting such researchers to tedious dual review. In general, regulation may be good in keeping controversy to a minimum, but it also has its price in slowing down daring attempts to find cures. If every great scientist in the past had to pass his proposals through two committees of peers, would we still have all our victories over disease?

The Dangers of Banning
Germany decided to legally ban, not regulate [gene therapy], and we can see the pitfalls in that approach. Germany passed legislation that shifted the onus of proof [of safety] to scientists and physicians. The default condition in Germany is to ban the procedures and use them only if evidence can be mounted proving them safe. It is a federal crime there to attempt any kind of genetic therapy.

Along this line, it is important to note that the European version of NBAC, meeting in The Hague, Holland, about the same time as the NBAC *Report* appeared, rejected even embryo twinning to increase pregnancy rates for in vitro fertilization. The committee decided that European women should be subjected to taking powerful, possibly cancerous drugs to stimulate superovulation rather than take the "risks" of twinning embryos. To allow such twinning, it said, would risk a slippery slope.

When the German law was passed, no conceivable therapeutic case could be imagined for genetic therapy. But knowledge accumulates fast, especially with the Human Genome Project, and now we know that there are some genetic diseases related to mitochondrial DNA that could be reduced by NST, especially from an embryo produced sexually and from which a nucleus is then transferred to an enucleated egg. To help German couples and children with these problems, a federal law must be overturned—not an easy undertaking. Far better not to build this obstacle into medicine and scientific research in the first place.

The Benefits of Regulation

Regulating NST rather than legislatively banning it could have many beneficial effects. For NST research, regulation would impose a single set of standards on all American clinics. Regulation would require some supervision of each NST experiment. Regulation could be supervised by a national oversight panel, such as the NIH peer-review panels that now judge applications for NIH funding. Regulation would allow public supervision, education, and control of publicly-funded NST research. Ideally, such regulation would have a sunset provision, allowing it to expire after a half-dozen years.

For the above reasons and others, I favor regulation of NST in infertility clinics, but only of those IVF clinics that want to do NST, for the first years of applying this technique. In addition to the above reasons, I think that humans are complex enough that we should go slowly to understand all the things that can go wrong.

CHAPTER 5

SHOULD HUMANS EVER BE CLONED?

ETHICAL ISSUES CONCERNING HUMAN CLONING

Wray Herbert, Jeffery L. Sheler, and Traci Watson

In the following article, Wray Herbert, Jeffery L. Sheler, and Traci Watson describe the chief ethical issues raised by the possibility of human cloning. The authors provide answers to a variety of frequently asked questions concerning the ethical and moral implications of human cloning. For instance, they explain that a human clone would not be an exact copy of the person who provided its DNA, but rather would be a sort of "delayed" identical twin, as much a separate individual as any twin is. They discuss the circumstances under which cloning a human might be deemed ethical, as well as the theoretical misuses of such technology. Finally, they consider the possible effects—both good and bad—of human cloning on family relationships, society, and religion. Herbert, Sheler, and Watson are all staff writers for *U.S. News & World Report*.

At first it was just plain startling. Word from Scotland in late February 1997 that a scientist named Ian Wilmut had succeeded in cloning an adult mammal—a feat long thought impossible—caught the imagination of even the most jaded technophobe. The laboratory process that produced Dolly, an unremarkable-looking sheep, theoretically would work for humans as well. A world of clones and drones, of *The Boys From Brazil* and *Multiplicity* [two science fiction movies about cloning], was suddenly within reach. It was science fiction come to life. And scary science fiction at that.

In the wake of Wilmut's shocker, governments scurried to formulate guidelines for the unknown, a future filled with mind-boggling possibilities. The Vatican called for a worldwide ban on human cloning. President Bill Clinton ordered a national commission to study the legal and ethical implications. Leaders in Europe, where most nations already prohibit human cloning, began examining the moral ramifications of cloning other species.

Like the splitting of the atom, the first space flight, and the discovery of "life" on Mars, Dolly's debut has generated a long list of difficult puzzles for scientists and politicians, philosophers and theologians. And at dinner tables and office coolers, in bars and on street corners, the development of wild scenarios spun from the birth of a simple sheep has only just begun. *U.S. News* sought answers from experts to the most intriguing and frequently asked questions.

The Possibilities of Human Cloning

Why would anyone want to clone a human being in the first place? The human cloning scenarios that ethicists ponder most frequently fall into two broad categories: 1) parents who want to clone a child, either to provide transplants for a dying child or to replace that child, and 2) adults who for a variety of reasons might want to clone themselves.

Many ethicists, however, believe that after the initial period of uproar, there won't be much interest in cloning humans. Making copies, they say, pales next to the wonder of creating a unique human being the old-fashioned way.

Could a human being be cloned today? What about other animals? It would take years of trial and error before cloning could be applied successfully to other mammals. For example, scientists will need to find out if the donor egg is best used when it is resting quietly or when it is growing.

Will it be possible to clone the dead? Perhaps, if the body is fresh, says Randall Prather, a cloning expert at the University of Missouri–Columbia. The cloning method used by Wilmut's lab requires fusing an egg cell with the cell containing the donor's DNA. And that means the donor cell must have an intact membrane around its DNA. The membrane starts to fall apart after death, as does DNA. But, yes, in theory at least it might be possible.

Can I set up my own cloning lab? Yes, but maybe you'd better think twice. All the necessary chemicals and equipment are easily available and relatively low-tech. But out-of-pocket costs would run $100,000 or more, and that doesn't cover the pay for a skilled developmental biologist. The lowest-priced of these scientists, straight out of graduate school, makes about $40,000 a year. If you tried to grow the cloned embryos to maturity, you'd encounter other difficulties. The Scottish team implanted 29 very young clones in 13 ewes, but only one grew into a live lamb. So if you plan to clone Fluffy, buy enough cat food for a host of surrogate mothers.

A Twin, Not a Copy

Would a cloned human be identical to the original? Identical genes don't produce identical people, as anyone acquainted with identical twins can tell you. In fact, twins are more alike than clones would be, since they have at least shared the uterine environment, are usually raised in the same family, and so forth. Parents could clone a second child who eerily resembled their first in appearance, but all the evidence suggests the two would have very different personalities. Twins separated at birth do sometimes share quirks of personality, but such quirks in a cloned son or daughter would be haunting reminders of the child who was lost—and the failure to re-create that child.

Even biologically, a clone would not be identical to the "master copy." The clone's cells, for example, would have energy-processing machinery (mitochondria) that came from the egg donor, not from the nucleus donor. But most of the physical differences between originals and copies wouldn't be detectable without a molecular-biology lab. The one possible exception is fertility. Wilmut and his coworkers are not sure that Dolly will be able to have lambs. They will try to find out once she's old enough to breed.

Will a cloned animal die sooner or have other problems because its DNA is older? Scientists don't know. For complex biological reasons, creating a clone from an older animal differs from breeding an older animal in the usual way. So clones of adults probably wouldn't risk the same birth defects as the offspring of older women, for example. But the age of the DNA used for the clone still might affect life span. The Scottish scientists will monitor how gracefully Dolly ages.

What if parents decided to clone a child in order to harvest organs? Most experts agree that it would be psychologically harmful if a child sensed he had been brought into the world simply as a commodity. But some parents already conceive second children with nonfatal bone marrow transplants in mind, and many ethicists do not oppose this. Cloning would increase the chances for a biological match from 25 percent to nearly 100 percent.

If cloned animals could be used as organ donors, we wouldn't have to worry about cloning twins for transplants. Pigs, for example, have organs similar in size to humans'. But the human immune system attacks and destroys tissue from other species. To get around that, the Connecticut biotech company Alexion Pharmaceuticals Inc. is trying to alter the pig's genetic codes to prevent rejection. If Alexion succeeds, it may be more efficient to mass-produce porcine organ donors by cloning than by current methods, in which researchers inject pig embryos with human genes and hope the genes get incorporated into the embryo's DNA.

Wouldn't it be strange for a cloned twin to be several years younger than his or her sibling? When the National Advisory Board on Ethics in Reproduction studied a different kind of cloning a few years ago, its members split on the issue of cloned twins separated in time. Some thought the children's individuality might be threatened, while others argued that identical twins manage to keep their individuality intact.

Redefining Family Relationships

John Robertson of the University of Texas raises several other issues worth pondering: What about the cloned child's sense of free will and parental expectations? Since the parents chose to duplicate their first child, will the clone feel obliged to follow in the older sibling's footsteps? Will the older child feel he has been duplicated because he was inadequate or because he is special? Will the two have a unique form of sibling rivalry, or a special bond? These are, of course, just special versions of questions that come up whenever a new child is introduced into a family.

Could a megalomaniac decide to achieve immortality by cloning an "heir"? Sure, and there are other situations where adults might be tempted to clone themselves. For example, a couple in which the man is infertile might opt to clone one of them rather than introduce an outsider's sperm. Or a single woman might choose to clone herself rather than involve a man in any way. In both cases, however, you would have adults raising children who are also their twins—a situation ethically indistinguishable from the megalomaniac cloning himself. On adult cloning, ethicists are more united in their discomfort. In fact, the same commission that was divided on the issue of twins was unanimous in its conclusion that cloning an adult's twin is "bizarre . . . narcissistic and ethically impoverished." What's more, the commission argued that the phenomenon would jeopardize our very sense of who's who in the world, especially in the family.

How would a human clone refer to the donor of its DNA? "Mom" is not right, because the woman or women who supplied the egg and the womb would more appropriately be called Mother. "Dad" isn't right, either. A traditional father supplies only half the DNA in an offspring. Judith Martin, etiquette's "Miss Manners," suggests, "Most honored sir or madame." Why? "One should always respect one's ancestors," she says, "regardless of what they did to bring one into the world."

That still leaves some linguistic confusion. Michael Agnes, editorial director of *Webster's New World Dictionary*, says that "clonee" may sound like a good term, but it's too ambiguous. Instead, he prefers "original" and "copy." And above all else, advises Agnes, "Don't use 'Xerox.'"

A scientist joked that cloning could make men superfluous. Is it true? Yes, theoretically. A woman who wanted to clone herself would not need a man. Besides her DNA, all she would require are an egg and a womb—her own or another woman's. A man who wanted to clone himself, on the other hand, would need to buy the egg and rent the womb—or find a very generous woman.

Implications for Society

What are the other implications of cloning for society? The gravest concern about the misuse of genetics isn't related to cloning directly, but to genetic engineering—the deliberate manipulation of genes to enhance human talents and create human beings according to certain specifications. But some ethicists also are concerned about the creation of a new (and stigmatized) social class: "the clones." Albert Jonsen of the University of Washington believes the confrontation could be comparable to what occurred in the 16th century, when Europeans were perplexed by the unfamiliar inhabitants of the New World and endlessly debated their status as humans.

Whose pockets will cloning enrich in the near future? Not Ian Wilmut's. He's a government employee and owns no stock in PPL Therapeutics, the British company that holds the rights to the cloning technology. On the other hand, PPL stands to make a lot of money. Also likely to cash in are pharmaceutical and agricultural companies and maybe even farmers. The biotech company Genzyme has already bred goats that are genetically engineered to give milk laced with valuable drugs. Wilmut and other scientists say it would be much easier to produce such animals with cloning than with today's methods. Stock breeders could clone champion dairy cows or the meatiest pigs.

Could cloning be criminally misused? If the technology to clone humans existed today, it would be almost impossible to prevent someone from cloning you without your knowledge or permission, says Philip Bereano, professor of technology and public policy at the University of Washington. Everyone gives off cells all the time—whenever we give a blood sample, for example, or visit the dentist—and those cells all contain one's full complement of DNA. What would be the goal of such "drive-by" cloning? Well, what if a woman were obsessed with having the child of an apathetic man? Or think of the commercial value of a dynasty-building athletic pedigree or a heavenly singing voice. Even though experience almost certainly shapes these talents as much as genetic gifts, the unscrupulous would be unlikely to be deterred.

Religious Attitudes

Is organized religion opposed to cloning? Many of the ethical issues being raised about cloning are based in theology. Concern for preserving human dignity and individual freedom, for example, is deeply rooted in religious and biblical principles. But until Wilmut's announcement, there had been surprisingly little theological discourse on the implications of cloning per se. The response so far from the religious community, while overwhelmingly negative, has been far from monolithic.

Roman Catholic, Protestant, and Jewish theologians all caution against applying the new technology to humans, but for varying reasons. Catholic opposition stems largely from the church's belief that "natural moral law" prohibits most kinds of tampering with human reproduction. A 1987 Vatican document, *Donum Vitae*, condemned cloning because it violates "the dignity both of human procreation and of the conjugal union."

Protestant theology, on the other hand, emphasizes the view that nature is "fallen" and subject to improvement. "Just because some-thing occurs naturally doesn't mean it's automatically good," explains Max Stackhouse of Princeton Theological Seminary. But while they tend to support using technology to fix flaws in nature, Protestant theologians say cloning of humans crosses the line. It places too much power in the hands of sinful humans, who, says philosophy Prof. David Fletcher of Wheaton College in Wheaton, Ill., are subject to committing "horrific abuses."

Judaism also tends to favor using technology to improve on nature's shortcomings, says Rabbi Richard Address of the Union of American Hebrew Congregations. But cloning humans, he says, "is an area where we cannot go. It violates the mystery of what it means to be human."

Doesn't cloning encroach on the Judeo-Christian view of God as the creator of life? Would a clone be considered a creature of God or of science? Many theologians worry about this. Cloning, at first glance, seems to be a usurpation of God's role as creator of humans "in his own image." The scientist, rather than God or chance, deter-mines the outcome. "Like Adam and Eve, we want to be like God, to be in control," says philosophy Prof. Kevin Wildes of Georgetown University. "The question is, what are the limits?"

But some theologians argue that cloning is not the same as creating life from scratch. The ingredients used are alive or contain the ele-ments of life, says Fletcher of Wheaton College. It is still only God, he says, who creates life.

Would a cloned person have its own soul? Most theologians agree with scientists that a human clone and its DNA donor would be sepa-rate and distinct persons. That means each would have his or her own body, mind, and soul.

Would cloning upset religious views about death, immortality, and even resurrection? Not really. Cloned or not, we all die. The clone that outlives its "parent"—or that is generated from the DNA of a dead person, if that were possible—would be a different person. It would not be a reincarnation or a resurrected version of the deceased. Cloning could be said to provide immortality, theologians say, only in the sense that, as in normal reproduction, one might be said to "live on" in the genetic traits passed to one's progeny.

JUDEO-CHRISTIAN OBJECTIONS TO HUMAN CLONING

Stephen G. Post

Stephen G. Post is an associate professor of bioethics at the Center for Biomedical Ethics at Case Western Reserve University in Cleveland, Ohio. In the following essay, he presents the Judeo-Christian case against cloning human beings. Cloning, Post claims, would eliminate the mystery of newness and unique individuality that should surround every child. He also cautions that cloning might threaten the physical or psychological health of children produced through this technology. Even more importantly, Post maintains, cloning demonstrates a lack of respect for God and for the union of sex, marriage, love, and procreation mandated by the Hebrew Bible and adopted by Christianity. Both the Jewish and the Christian traditions, according to Post, place strong emphasis on monogamy and the family—traditions that human cloning could unravel.

The very idea of cloning tends to focus on the physiological substrate [biological aspects], not on the journey of life and our responses to it.

For purposes of discussion, I will assume that the cloning of humans is technologically possible. This supposition raises Albert Einstein's concern: "Perfection of means and confusion of ends seems to characterize our age." Public reaction to human cloning has been strongly negative, although without much clear articulation as to why. My task is the Socratic one of helping to make explicit what is implicit in this uneasiness.

Some extremely hypothetical scenarios might be raised as if to justify human cloning. One might speculate, for example: If environmental toxins or pathogens should result in massive human infertility, human cloning might be imperative for species survival. But in fact recent claims about increasing male infertility worldwide have been found to be false. Some apologists for human cloning will insist on other strained "What if's." "What if" parents want to replace a dead child with an image of that child? "What if" we can enhance the human condition by cloning the "best" among us? I shall offer seven

Excerpted from "The Judeo-Christian Case Against Cloning," by Stephen G. Post, *America,* June 21, 1997. Reprinted with permission from the author.

unhypothetical criticisms of human cloning, but in no particular priority. The final criticism, however, is the chief one to which all else serves as preamble.

Unique Lives

1. The Newness of Life. Although human cloning, if possible, is surely a novelty, it does not corner the market on newness. For millennia mothers and fathers have marveled at the newness of form in their newborns. I have watched newness unfold in our own two children, wonderful blends of the Amerasian variety. True, there probably is, as psychoanalyst Sigmund Freud argued, a certain narcissism in parental love, for we do see our own form partly reflected in the child, but, importantly, never entirely so. Sameness is dull, and as the French say, *Vive la différence.* It is possible that underlying the mystery of this newness of form is a creative wisdom that we humans will never quite equal.

This concern with the newness of each human form (identical twins are new genetic combinations as well) is not itself new. The scholar of constitutional law Laurence Tribe pointed out in 1978, for example, that human cloning could "alter the very meaning of humanity." Specifically, the cloned person would be "denied a sense of uniqueness." Let us remember that there is no strong analogy between human cloning and natural identical twinning, for in the latter case there is still the blessing of newness in the newborns, though they be two or more. While identical twins do occur naturally and are unique persons, this does not justify the temptation to impose external sameness more widely.

Sidney Callahan, a thoughtful psychologist, argues that the random fusion of a couple's genetic heritage "gives enough distance to allow the child also to be seen as a separate other," and she adds that the egoistic intent to deny uniqueness is wrong because ultimately depriving. By having a different form from that of either parent, I am visually a separate creature, and this contributes to the moral purpose of not reducing me to a mere copy utterly controlled by the purposes of a mother or father.

Perhaps human clones will not look exactly alike anyway, given the complex factors influencing genetic imprinting, as well as environmental factors affecting gene expression. But they will look more or less the same, rather than more or less different.

Surely no scientist would doubt that genetic diversity produced by procreation between a man and a woman will always be preferable to cloning, because procreation reduces the possibility for species annihilation through particular diseases or pathogens. Even in the absence of such pathogens, cloning means the loss of what geneticists describe as the additional hybrid vigor of new genetic combinations.

2. Making Males Reproductively Obsolete. Cloning requires human eggs, nuclei and uteri, all of which can be supplied by women. This makes males reproductively obsolete. This does not quite measure up to Shulamith Firestone's notion that women will only be able to free themselves from patriarchy through the eventual development of the artificial womb, but of course, with no men available, patriarchy ends—period.

Cloning, in the words of Richard McCormick, S.J., "would involve removing insemination and fertilization from the marriage relationship, and it would also remove one of the partners from the entire process." Well, removal of social fatherhood is already a *fait accompli* in a culture of illegitimacy chic, and one to which some fertility clinics already marvelously contribute through artificial insemination by donor for single women. Removing male impregnators from the procreative dyad would simply drive the nail into the coffin of fatherhood, unless one thinks that biological and social fatherhood are utterly disconnected. Social fatherhood would still be possible in a world of clones, but this will lack the feature of participation in a continued biological lineage that seems to strengthen social fatherhood in general.

Issues of Power and Control

3. Under My Thumb: Cookie Cutters and Power. It is impossible to separate human cloning from concerns about power. There is the power of one generation over the external form of another, imposing the vicissitudes of one generation's fleeting image of the good upon the nature and destiny of the next. One need only peruse the innumerable texts on eugenics written by American geneticists in the 1920s to understand the arrogance of such visions.

One generation always influences the next in various ways, of course. But when one generation can, by the power of genetics, in the words of C.S. Lewis, "make its descendants what it pleases, all men who live after it are the patients of that power." What might our medicalized culture's images of human perfection become? In Lewis' words again, "For the power of Man to make himself what he pleases means, as we have seen, the power of some men to make other men what they please."

A certain amount of negative eugenics by prenatal testing and selective abortion is already established in American obstetrics. Cloning extends this power from the negative to the positive, and it is therefore even more foreboding.

This concern with overcontrol and overpower may be overstated because the relationship between genotype and realized social role remains highly obscure. Social role seems to be arrived at as much through luck and perseverance as anything else, although some innate capacities exist as genetically informed baselines.

4. Born to Be Harvested. One hears regularly that human clones would make good organ donors. But we ought not to presume that anyone wishes to give away body parts. The assumption that the clone would choose to give body parts is completely unfounded. Forcing such a harvest would reduce the clone to a mere object for the use of others. A human person is an individual substance of a rational nature not to be treated as object, even if for one's own narcissistic gratification, let alone to procure organs. I have never been convinced that there are any ethical duties to donate organs.

5. *The Problem of Mishaps.* Dolly the celebrated ewe represents one success out of 277 embryos, nine of which were implanted. Only Dolly survived. While I do not wish to address here the issue of the moral status of the entity within the womb, suffice it to note that in this country there are many who would consider proposed research to clone humans as far too risky with regard to induced genetic defects. Embryo research in general is a matter of serious moral debate in the United States, and cloning will simply bring this to a head.

As one recent British expert on fertility studies writes, "Many of the animal clones that have been produced show serious developmental abnormalities, and, apart from ethical considerations, doctors would not run the medico-legal risks involved."

6. *Sources of the Self.* Presumably no one needs to be reminded that the self is formed by experience, environment and nurture. From a moral perspective, images of human goodness are largely virtue-based and therefore characterological. Aristotle and Thomas Aquinas believed that a good life is one in which, at one's last breath, one has a sense of integrity and meaning. Classically the shaping of human fulfillment has generally been a matter of negotiating with frailty and suffering through perseverance in order to build character. It is not the earthen vessels, but the treasure within them that counts. A self is not so much a genotype as a life journey. Martin Luther King Jr. was getting at this when he said that the content of character is more important than the color of skin.

The very idea of cloning tends to focus images of the good self on the physiological substrate, not on the journey of life and our responses to it, some of them compensations to purported "imperfections" in the vessel. The idea of the designer baby will emerge, as though external form is as important as the inner self.

God's Creations

7. *Respect for Nature and Nature's God.* In the words of Jewish bioethicist Fred Rosner, cloning goes so far in violating the structure of nature that it can be considered as "encroaching on the Creator's domain." Is the union of sex, marriage, love and procreation something to dismiss lightly?

Marriage is the union of female and male that alone allows for procreation in which children can benefit developmentally from both a mother and father. In the Gospel of Mark, Jesus draws on ancient Jewish teachings when he asserts, "Therefore what God has joined together, let no man separate." Regardless of the degree of extendedness in any family, there remains the core nucleus: wife, husband and children. Yet the nucleus can be split by various cultural forces (e.g., infidelity as interesting, illegitimacy as chic), poverty, patriarchal violence and now cloning.

A cursory study of the Hebrew Bible shows the exuberant and immensely powerful statements of Genesis 1, in which a purposeful, ordering God pronounces that all stages of creation are "good." The text proclaims, "So God created humankind in his image, in the image of God he created them, male and female he created them" (Gen. 1:27). This God commands the couple, each equally in God's likeness, to "be fruitful and multiply." The divine prototype was thus established at the very outset of the Hebrew Bible: "Therefore a man leaves his father and his mother and clings to his wife, and they become one flesh" (Gen. 2:24).

The dominant theme of Genesis 1 is creative intention. God creates, and what is created procreates, thereby ensuring the continued presence of God's creation. The creation of man and woman is good in part because it will endure. Catholic natural law ethicists and Protestant proponents of "orders of creation" alike find divine will and principle in the passages of Genesis 1.

The Sacredness of the Family

A major study on the family by the Christian ethicist Max Stackhouse suggests that just as the pre-Socratic philosophers discovered still valid truths about geometry, so the biblical authors of Chapters One and Two of Genesis "saw something of the basic design, purpose, and context of life that transcends every sociohistorical epoch." Specifically, this design includes "fidelity in communion" between male and female oriented toward "generativity" and an enduring family the precise social details of which are worked out in the context of political economies.

Christianity appropriated the Hebrew Bible and had its origin in a Jew from Nazareth and his Jewish followers. The basic contours of Christian thought on marriage and family therefore owe a great deal to Judaism. These Hebraic roots that shape the words of Jesus stand within Malachi's prophetic tradition of emphasis on inviolable monogamy. In Mark 10:2–12 we read:

> The Pharisees approached and asked, "Is it lawful for a husband to divorce his wife?" They were testing him. He said to them in reply, "What did Moses command you?" They replied, "Moses permitted him to write a bill of divorce and dismiss her." But Jesus told them, "Because of the hardness of your hearts he wrote you this commandment. But from the begin-

ning of creation, 'God made them male and female. For this reason a man shall leave his father and mother (and be joined to his wife), and the two shall become one flesh.' So they are no longer two but one flesh. Therefore what God has joined together, no human being must separate." In the house the disciples again questioned him about this. He said to them, "Whoever divorces his wife and marries another commits adultery against her; and if she divorces her husband and marries another, she commits adultery."

Here Jesus quotes Gen. 1:27 ("God made them male and female") and Gen. 2:24 ("the two shall become one flesh").

Christians side with the deep wisdom of the teachings of Jesus, manifest in a thoughtful respect for the laws of nature that reflect the word of God. Christians simply cannot and must not underestimate the threat of human cloning to unravel what is both naturally and eternally good.

CLONING COULD HALT HUMAN EVOLUTION

Michael Mautner

Michael Mautner warns that if cloning ever becomes the preferred method of human reproduction, it could feasibly halt human evolution because it would eliminate the source of genetic diversity that sexual reproduction provides. A child produced through the joining of a sperm cell and an egg receives half its genes from its father and half from its mother, whereas a clone would receive all of its genes from a single individual, he explains. This loss of diversity, Mautner says, could create a situation where no one possesses genetic resistance to certain diseases, leaving all of humanity vulnerable to extermination by an epidemic. Mautner argues that humans need to evolve further to deal with potential environmental crises on Earth and the diverse environments that they may encounter during space exploration. Cloning, he concludes, could put an end to this essential process of ongoing evolution. Mautner is a chemistry professor at the University of Canterbury at Christchurch, New Zealand, and has written several articles for the *Futurist*.

Cloning is not only less fun than sex, it would freeze evolution and destroy our chances for survival in the future.

The cloning of the first mammal brings the prospects of human cloning closer to reality. Now the public should ponder the implications. Among these, the most important is the effect on our future evolution.

The Need for Diversity

Cloning will be attractive because of some medical uses. Genetic replicas of geniuses might also benefit society. On the other hand, ruthless and egocentric despots may replicate themselves millions of times over. Cloning on a large scale would also reduce biological diversity, and the entire human species could be wiped out by some new epidemic to which a genetically uniform population was susceptible.

Excerpted from "Will Cloning End Human Evolution?" by Michael Mautner, *The Futurist*, November/December 1997. Reprinted with permission from the World Future Society, 7910 Woodmont Ave., Suite 450, Bethesda, MD 20814; ph. (301) 656-8274, fax (301) 951-0394; www.wfs.org.

Beyond these important but obvious results, cloning raises problems that go to the core of human existence and purpose. One important fact to recognize is that cloning is asexual reproduction. It therefore bypasses both the biological benefits of normal reproduction and the emotional, psychological, and social aspects that surround it: courtship, love, marriage, family structure. Even more importantly, if cloning became the main mode of reproduction, human evolution would stop in its tracks.

In sexual reproduction, some of the genetic material from each parent undergoes mutations that can lead to entirely new biological properties. Vast numbers of individual combinations become possible, and the requirements of survival—and choices of partners by the opposite sex—then gradually select which features will be passed on to the following generations.

Cloning will, in contrast, reproduce the same genetic makeup of an existing individual. There is no room for new traits to arise by mutation and no room for desirable features to compete and win by an appeal to the judgment of the opposite sex. The result: Human evolution is halted.

Evolving in New Worlds

Is it necessary for the human species to evolve further? Absolutely! We are certainly far from achieving perfection. We are prone to diseases, and the capacity of our intelligence is limited. Most importantly, human survival will depend on our ability to adapt to environments beyond Earth—that is, in the rich new worlds of outer space.

Some people question whether we can save ourselves from manmade environmental disasters on Earth, whose resources are already pressured by human population growth. And limiting the population to one planet puts us at risk of extinction from all-out nuclear or biological warfare, climate change, and catastrophic meteorite impacts.

Humanity could vastly expand its chances for survival by moving into space, where we would encounter worlds with diverse environments. To live in space, we will have to increase our tolerance to radiation, to extremes of heat and cold, and to vacuum. We will also need more intelligence to construct habitats. Our social skills will need to advance so that billions of humans can work together in the grand projects that will be needed.

If we are to expand into space, we surely cannot freeze human evolution. The natural (and possibly designed) mechanisms of evolution must therefore be allowed to continue.

Preserving Life

Socially, the relations between the sexes underlie most aspects of human behavior. The rituals of dating, mating, and marriage and the family structures that surround sexual reproduction are the most basic emotional and social factors that make our lives human. Without the satisfactions of love and sex, of dating and of families, will cloned generations even care to propagate further?

Cloning therefore raises fundamental questions about the human future: Have we arrived yet at perfection? Where should we aim future human evolution? What is the ultimate human purpose? The prospect of human cloning means that these once-philosophical questions have become urgent practical issues.

As living beings, our primary human purpose is to safeguard, propagate, and advance life. This objective must guide our ethical judgments, including those on cloning.

Our best guide to this purpose is the love of life common to most humans, which is therefore reflected in our communal judgment. All individuals who sustain the present and build the future should have the right to participate equally in these basic decisions. Our shared future may be best secured by the practice of debating and voting on such biotechnology issues in an informed "biodemocracy."

WHY NOT CLONE HUMANS?

Robert Wachbroit

In the following selection, Robert Wachbroit takes issue with some of the primary ethical objections against human cloning. For example, he notes that many experts fear that cloned children will face unrealistic expectations that they will have the same talents, interests, or career goals as the persons from whom they were cloned rather than developing their own individual personalities. While this problem might indeed occur, Wachbroit argues, it would not be due to the nature of cloning itself. He points out that many parents already have this kind of unrealistic expectation for their offspring. Bad parenting can occur regardless of whether a child is produced through normal reproduction or human cloning, Wachbroit maintains, and should not be used as a reason to prevent the development of cloning. Wachbroit is a research scholar at the Institute for Philosophy and Public Policy, part of the School of Public Affairs at the University of Maryland in College Park.

The successful cloning of an adult sheep, announced in Scotland in February 1997, is one of the most dramatic recent examples of a scientific discovery becoming a public issue. During the last few months, various commentators—scientists and theologians, physicians and legal experts, talk-radio hosts and editorial writers—have been busily responding to the news, some calming fears, others raising alarms about the prospect of cloning a human being. At the request of the President, the National Bioethics Advisory Commission (NBAC) held hearings and prepared a report on the religious, ethical, and legal issues surrounding human cloning. While declining to call for a permanent ban on the practice, the Commission recommended a moratorium on efforts to clone human beings, and emphasized the importance of further public deliberation on the subject.

An interesting tension is at work in the NBAC report. Commission members were well aware of "the widespread public discomfort, even revulsion, about cloning human beings." Perhaps recalling the images of Dolly the ewe that were featured on the covers of national news

Excerpted from "Genetic Encores: The Ethics of Human Cloning," by Robert Wachbroit, *Report from the Institute for Philosophy and Public Policy*, Fall 1997. Reprinted with permission.

magazines, they noted that "the impact of these most recent developments on our national psyche has been quite remarkable." Accordingly, they felt that one of their tasks was to articulate, as fully and sympathetically as possible, the range of concerns that the prospect of human cloning had elicited.

Genetic Determinism

Yet it seems clear that some of these concerns, at least, are based on false beliefs about genetic influence and the nature of the individuals that would be produced through cloning. Consider, for instance, the fear that a clone would not be an "individual" but merely a "carbon copy" of someone else—an automaton of the sort familiar from science fiction. As many scientists have pointed out, a clone would not in fact be an identical *copy*, but more like a delayed identical *twin*. And just as identical twins are two separate people—biologically, psychologically, morally and legally, though not genetically—so, too, a clone would be a separate person from her non-contemporaneous twin. To think otherwise is to embrace a belief in genetic determinism—the view that genes determine everything about us, and that environmental factors or the random events in human development are insignificant.

The overwhelming scientific consensus is that genetic determinism is false. In coming to understand the ways in which genes operate, biologists have also become aware of the myriad ways in which the environment affects their "expression." The genetic contribution to the simplest physical traits, such as height and hair color, is significantly mediated by environmental factors (and possibly by stochastic [random] events as well). And the genetic contribution to the traits we value most deeply, from intelligence to compassion, is conceded by even the most enthusiastic genetic researchers to be limited and indirect.

It is difficult to gauge the extent to which "repugnance" toward cloning generally rests on a belief in genetic determinism. Hoping to account for the fact that people "instinctively recoil" from the prospect of cloning, James Q. Wilson wrote, "There is a natural sentiment that is offended by the mental picture of identical babies being produced in some biological factory." Which raises the question: once people learn that this picture is mere science fiction, does the offense that cloning presents to "natural sentiment" attenuate, or even disappear? Jean Bethke Elshtain cited the nightmare scenarios of "the man and woman on the street," who imagine a future populated by "a veritable army of Hitlers, ruthless and remorseless bigots who kept reproducing themselves until they had finished what the historic Hitler failed to do: annihilate us." What happens, though, to the "pity and terror" evoked by the topic of cloning when such scenarios are deprived (as they deserve to be) of all credibility?

Richard Lewontin has argued that the critics' fears—or at least, those fears that merit consideration in formulating public policy—dissolve once genetic determinism is refuted. He criticizes the NBAC report for excessive deference to opponents of human cloning, and calls for greater public education on the scientific issues. (The Commission in fact makes the same recommendation, but Lewontin seems unimpressed.) Yet even if a public education campaign succeeded in eliminating the most egregious misconceptions about genetic influence, that wouldn't settle the matter. People might continue to express concerns about the interests and rights of human clones, about the social and moral consequences of the cloning process, and about the possible motivations for creating children in this way.

The Right to an Open Future

One set of ethical concerns about human clones involves the risks and uncertainties associated with the current state of cloning technology. This technology has not yet been tested with human subjects, and scientists cannot rule out the possibility of mutation or other biological damage. Accordingly, the NBAC report concluded that "at this time, it is morally unacceptable for anyone in the public or private sector, whether in a research or clinical setting, to attempt to create a child using somatic cell nuclear transfer cloning." Such efforts, it said, would pose "unacceptable risks to the fetus and/or potential child."

The ethical issues of greatest importance in the cloning debate, however, do not involve possible failures of cloning technology, but rather the consequences of its success. Assuming that scientists were able to clone human beings without incurring the risks mentioned above, what concerns might there be about the welfare of clones?

Some opponents of cloning believe that such individuals would be wronged in morally significant ways. Many of these wrongs involve the denial of what Joel Feinberg has called "the right to an open future." For example, a child might be constantly compared to the adult from whom he was cloned, and thereby burdened with oppressive expectations. Even worse, the parents might actually limit the child's opportunities for growth and development: a child cloned from a basketball player, for instance, might be denied any educational opportunities that were not in line with a career in basketball. Finally, regardless of his parents' conduct or attitudes, a child might be burdened by the *thought* that he is a copy and not an "original." The child's sense of self-worth or individuality or dignity, so some have argued, would thus be difficult to sustain.

How should we respond to these concerns? On the one hand, the existence of a right to an open future has a strong intuitive appeal. We are troubled by parents who radically constrict their children's possibilities for growth and development. Obviously, we would condemn a cloning parent for crushing a child with oppressive expectations, just as we might condemn fundamentalist parents for utterly isolating their children from the modern world, or the parents of twins for inflicting matching wardrobes and rhyming names. But this is not enough to sustain an objection to cloning itself. Unless the claim is that cloned parents cannot help but be oppressive, we would have cause to say they had wronged their children only because of their subsequent, and avoidable, sins of bad parenting—not because they had chosen to create the child in the first place. (The possible reasons for making this choice will be discussed below.)

We must also remember that children are often born in the midst of all sorts of hopes and expectations; the idea that there is a special burden associated with the thought "There is someone who is genetically just like me" is necessarily speculative. Moreover, given the falsity of genetic determinism, any conclusions a child might draw from observing the person from whom he was cloned would be uncertain at best. His knowledge of his future would differ only in degree from what many children already know once they begin to learn parts of their family's (medical) history. Some of us knew that we would be bald, or to what diseases we might be susceptible. To be sure, the cloned individual might know more about what he or she could become. But because our knowledge of the effect of environment on development is so incomplete, the clone would certainly be in for some surprises.

Finally, even if we were convinced that clones are likely to suffer particular burdens, that would not be enough to show that it is wrong to create them. The child of a poor family can be expected to suffer specific hardships and burdens, but we don't thereby conclude that such children shouldn't be born. Despite the hardships, poor children can experience parental love and many of the joys of being alive: the deprivations of poverty, however painful, are not decisive. More generally, no one's life is entirely free of some difficulties or burdens. In order for these considerations to have decisive weight, we have to be able to say that life doesn't offer any compensating benefits. Concerns expressed about the welfare of human clones do not appear to justify such a bleak assessment. Most such children can be expected to have lives well worth living; many of the imagined harms are no worse than those faced by children acceptably produced by more conventional means. If there is something deeply objectionable about cloning, it is more likely to be found by examining implications of the cloning process itself, or the reasons people might have for availing themselves of it.

Cloning and Other Technologies

Human cloning falls conceptually between two other technologies. At one end we have the assisted reproductive technologies, such as in vitro fertilization, whose primary purpose is to enable couples to produce a child with whom they have a biological connection. At the other end we have the emerging technologies of genetic engineering—specifically, gene transplantation technologies—whose primary purpose is to produce a child that has certain traits. Many proponents of cloning see it as part of the first technology: cloning is just another way of providing a couple with a biological child they might otherwise be unable to have. Since this goal and these other technologies are acceptable, cloning should be acceptable as well. On the other hand, many opponents of cloning see it as part of the second technology: even though cloning is a transplantation of an entire nucleus and not of specific genes, it is nevertheless an attempt to produce a child with certain traits. The deep misgivings we may have about the genetic manipulation of offspring should apply to cloning as well.

The debate cannot be resolved, however, simply by determining which technology to assimilate cloning to. For example, some opponents of human cloning see it as continuous with assisted reproductive technologies; but since they find those technologies objectionable as well, the assimilation does not indicate approval. Rather than argue for grouping cloning with one technology or another, I wish to suggest that we can best understand the significance of the cloning process by comparing it with these other technologies, and thus broadening the debate.

Cloning's Effect on Family Relationships

To see what can be learned from such a comparative approach, let us consider a central argument that has been made against cloning—that it undermines the structure of the family by making identities and lineages unclear. On the one hand, the relationship between an adult and the child cloned from her could be described as that between a parent and offspring. Indeed, some commentators have called cloning "asexual reproduction," which clearly suggests that cloning is a way of generating *descendants*. The clone, on this view, has only one biological parent. On the other hand, from the point of view of genetics, the clone is a *sibling*, so that cloning is more accurately described as "delayed twinning" rather than as asexual reproduction. The clone, on this view, has two biological parents, not one—they are the same parents as those of the person from whom that individual was cloned.

Cloning thus results in ambiguities. Is the clone an offspring or a sibling? Does the clone have one biological parent or two? The moral significance of these ambiguities lies in the fact that in many societies, including our own, lineage identifies responsibilities. Typically, the parent, not the sibling, is responsible for the child. But if no one is unambiguously the parent, so the worry might go, who is responsible for the clone? Insofar as social identity is based on biological ties, won't this identity be blurred or confounded?

Some assisted reproductive technologies have raised similar questions about lineage and identity. An anonymous sperm donor is thought to have no parental obligations towards his biological child. A surrogate mother may be required to relinquish all parental claims to the child she bears. In these cases, the social and legal determination of "who is the parent" may appear to proceed in defiance of profound biological facts, and to subvert attachments that we as a society are ordinarily committed to upholding. Thus, while the *aim* of assisted reproductive technologies is to allow people to produce or raise a child to whom they are biologically connected, such technologies may also involve the creation of social ties that are permitted to override biological ones.

In the case of cloning, however, ambiguous lineages would seem to be less problematic, precisely because no one is being asked to relinquish a claim on a child to whom he or she might otherwise acknowledge a biological connection. What, then, are the critics afraid of? It does not seem plausible that someone would have herself cloned and then hand the child over to her parents, saying, "You take care of her! She's *your* daughter!" Nor is it likely that, if the cloned individual did raise the child, she would suddenly refuse to pay for college on the grounds that this was not a sister's responsibility. Of course, policy-makers should address any confusion in the social or legal assignment of responsibility resulting from cloning. But there are reasons to think that this would be *less* difficult than in the case of other reproductive technologies.

Cloning and Genetic Engineering

Similarly, when we compare cloning with genetic engineering, cloning may prove to be the less troubling of the two technologies. This is true even though the dark futures to which they are often alleged to lead are broadly alike. For example, a recent *Washington Post* article examined fears that the development of genetic enhancement technologies might "create a market in preferred physical traits." The reporter asked, "Might it lead to a society of DNA haves and have-nots, and the creation of a new underclass of people unable to keep up with the genetically fortified Joneses?" Similarly, a member of the National Bioethics Advisory Commission expressed concern that cloning might become "almost a preferred practice," taking its place "on the continuum of providing the best for your child." As a consequence, parents who chose to "play the lottery of old-fashioned reproduction would be considered irresponsible."

Such fears, however, seem more warranted with respect to genetic engineering than to cloning. By offering some people—in all probability, members of the upper classes—the opportunity to acquire desired traits through genetic manipulation, genetic engineering could bring about a biological reinforcement (or accentuation) of existing social divisions. It is hard enough already for disadvantaged children to compete with their more affluent counterparts, given the material resources and intellectual opportunities that are often available only to children of privilege. This unfairness would almost certainly be compounded if genetic manipulation came into the picture. In contrast, cloning does not bring about "improvements" in the genome: it is, rather, a way of *duplicating* the genome—with all its imperfections. It wouldn't enable certain groups of people to keep getting better and better along some valued dimension.

To some critics, admittedly, this difference will not seem terribly important. Theologian Gilbert Meilaender, Jr., objects to cloning on the grounds that children created through this technology would be "designed as a product" rather than "welcomed as a gift." The fact that the design process would be more selective and nuanced in the case of genetic engineering would, from this perspective, have no moral significance. To the extent that this objection reflects a concern about the commodification of human life, we can address it in part when we consider people's reasons for engaging in cloning.

Reasons for Cloning

This final area of contention in the cloning debate is as much psychological as it is scientific or philosophical. If human cloning technology were safe and widely available, what use would people make of it? What reasons would they have to engage in cloning?

In its report to the President, the Commission imagined a few situations in which people might avail themselves of cloning. In one scenario, a husband and wife who wish to have children are both carriers of a lethal recessive gene:

> Rather than risk the one in four chance of conceiving a child who will suffer a short and painful existence, the couple considers the alternatives: to forgo rearing children; to adopt; to use prenatal diagnosis and selective abortion; to use donor gametes free of the recessive trait; or to use the cells of one of the adults and attempt to clone a child. To avoid donor gametes and selective abortion, while maintaining a genetic tie to their child, they opt for cloning.

In another scenario, the parents of a terminally ill child are told that only a bone marrow transplant can save the child's life. "With no other donor available, the parents attempt to clone a human being from the cells of the dying child. If successful, the new child will be a perfect match for bone marrow transplant, and can be used as a donor without significant risk or discomfort. The net result: two healthy children, loved by their parents, who happen [sic] to be identical twins of different ages."

The Commission was particularly impressed by the second example. That scenario, said the NBAC report, "makes what is probably the strongest possible case for cloning a human being, as it demonstrates how this technology could be used for lifesaving purposes." Indeed, the report suggests that it would be a "tragedy" to allow "the sick child to die because of a moral or political objection to such cloning." Nevertheless, we should note that many people would be morally uneasy about the use of a minor as a donor, regardless of whether the child were a result of cloning. Even if this unease is justifiably overridden by other concerns, the "transplant scenario" may not present a more compelling case for cloning than that of the infertile couple desperately seeking a biological child.

Most critics, in fact, decline to engage the specifics of such tragic (and presumably rare) situations. Instead, they bolster their case by imagining very different scenarios. Potential users of the technology, they suggest, are narcissists or control freaks—people who will regard their children not as free, original selves but as products intended to meet more or less rigid specifications. Even if such people are not genetic determinists, their recourse to cloning will indicate a desire to exert all possible influence over what "kind" of child they produce.

The critics' alarm at this prospect has in part to do, as we have seen, with concerns about the psychological burdens such a desire would impose on the clone. But it also reflects a broader concern about the values expressed, and promoted, by a society's reproductive policies. Critics argue that a society that enables people to clone themselves thereby endorses the most narcissistic reason for having children—to perpetuate oneself through a genetic encore. The demonstrable falsity of genetic determinism may detract little, if at all, from the strength of this motive. Whether or not clones will have a grievance against their parents for producing them with this motivation, the societal indulgence of that motivation is improper and harmful.

It can be argued, however, that the critics have simply misunderstood the social meaning of a policy that would permit people to clone themselves even in the absence of the heartrending exigencies described in the NBAC report. This country has developed a strong commitment to reproductive autonomy. (This commitment emerged in response to the dismal history of eugenics—the very history that is sometimes invoked to support restrictions on cloning.) With the exception of practices that risk coercion and exploitation—notably baby-selling and commercial surrogacy—we do not interfere with people's freedom to create and acquire children by almost any means, for almost any reason. This policy does not reflect a dogmatic libertarianism. Rather, it recognizes the extraordinary personal importance and private character of reproductive decisions, even those with significant social repercussions.

Our willingness to sustain such a policy also reflects a recognition of the moral complexities of parenting. For example, we know that the motives people have for bringing a child into the world do not necessarily determine the manner in which they raise him. Even when parents start out as narcissists, the experience of childrearing will sometimes transform their initial impulses, making them caring, respectful, and even self-sacrificing. Seeing their child grow and develop, they learn that she is not merely an extension of themselves. Of course, some parents never make this discovery; others, having done so, never forgive their children for it. The pace and extent of moral development among parents (no less than among children) is infinitely variable. Still, we are justified in saying that those who engage in cloning will not, by virtue of this fact, be immune to the transformative effects of parenthood—even if it is the case (and it won't always be) that they begin with more problematic motives than those of parents who engage in the "genetic lottery."

Moreover, the nature of parental motivation is itself more complex than the critics often allow. Though we can agree that narcissism is a vice not to be encouraged, we lack a clear notion of where pride in one's children ends and narcissism begins. When, for example, is it unseemly to bask in the reflected glory of a child's achievements? Imagine a champion gymnast who takes delight in her daughter's athletic prowess. Now imagine that the child was actually cloned from one of the gymnast's somatic cells. Would we have to revise our moral assessment of her pleasure in her daughter's success? Or suppose a man wanted to be cloned and to give his child opportunities he himself had never enjoyed. And suppose that, rightly or wrongly, the man took the child's success as a measure of his own untapped potential— an indication of the flourishing life he might have had. Is *this* sentiment blamable? And is it all that different from what many natural parents feel?

Why Cloning Will Remain Rare

Until recently, there were few ethical, social, or legal discussions about human cloning via nuclear transplantation, since the scientific consensus was that such a procedure was not biologically possible. With the appearance of Dolly, the situation has changed. But although it now seems more likely that human cloning will become feasible, we may doubt that the practice will come into widespread use.

I suspect it will not, but my reasons will not offer much comfort to the critics of cloning. While the technology for nuclear transplantation advances, other technologies—notably the technology of genetic engineering—will be progressing as well. Human genetic engineering will be applicable to a wide variety of traits; it will be more powerful than cloning, and hence more attractive to more people. It will also, as I have suggested, raise more troubling questions than the prospect of cloning has thus far.

The Medical Benefits of Human Cloning

The Human Cloning Foundation

The Human Cloning Foundation of Atlanta, Georgia, strongly supports the idea of human cloning and related technologies. The foundation is headed by Richard Seed, a physicist who created great controversy in December 1997 by announcing that he planned to try to clone humans soon. In the following selection, the foundation contends that human cloning or research that involves cloning human embryos could have important medical benefits, such as new treatments for infertility, providing tissues and organs for transplantation, and finding a cure for cancer. According to the foundation, many people would support such uses of human cloning as ethical and beneficial.

There are many ways in which human cloning is expected to benefit mankind. Below is a list of ways that it is expected to help people. This list is far from complete.

The Medical Benefits of Cloning Research

• Human cloning technology could be used to reverse heart attacks. Scientists believe that they may be able to treat heart attack victims by cloning their healthy heart cells and injecting them into the areas of the heart that have been damaged. Heart disease is the number one killer in the United States and several other industrialized countries.

• There has been a breakthrough with human stem cells. Embryonic stem cells can be grown to produce organs or tissues to repair or replace damaged ones. Skin for burn victims, brain cells for the brain-damaged, spinal cord cells for quadriplegics and paraplegics, hearts, lungs, livers, and kidneys could be produced. By combining this technology with human cloning technology, it may be possible to produce needed tissue for suffering people that will be free of rejection by their immune systems. Conditions such as Alzheimer's disease, Parkinson's disease, diabetes, heart failure, degenerative joint disease, and other problems may be made curable if human cloning and its technology are not banned.

Excerpted from "The Benefits of Human Cloning," by The Human Cloning Foundation (1998). Article available at www.humancloning.org.

• Infertility: With cloning, infertile couples could have children. Despite getting a fair amount of publicity in the news, current treatments for infertility, in terms of percentages, are not very successful. One estimate is that current infertility treatments are less than 10 percent successful. Couples go through physically and emotionally painful procedures for a small chance of having children. Many couples run out of time and money without successfully having children. Human cloning could make it possible for many more infertile couples to have children than ever before possible.

• Plastic, reconstructive, and cosmetic surgery: Because of human cloning and its technology, the days of silicone breast implants and other cosmetic procedures that may cause immune disease should soon be over. With the new technology, instead of using materials foreign to the body for such procedures, doctors will be able to manufacture bone, fat, connective tissue, or cartilage that matches the patient's tissues exactly. Anyone will be able to have their appearance altered to their satisfaction without the leaking of silicone gel into their bodies or the other problems that occur with present-day plastic surgery. Victims of terrible accidents that deform the face should now be able to have their features repaired with new, safer technology. Limbs for amputees may be able to be regenerated.

• Breast implants: Most people are aware of the breast implant fiasco in which hundreds of thousands of women received silicone breast implants for cosmetic reasons. Many came to believe that the implants were making them ill with diseases of their immune systems. With human cloning and its technology, breast augmentation and other forms of cosmetic surgery could be done with implants that would not be any different from the person's normal tissues.

• Defective genes: The average person carries 8 defective genes inside them. These defective genes allow people to become sick when they would otherwise remain healthy. With human cloning and its technology, it may be possible to ensure that we no longer suffer because of our defective genes.

• Down's syndrome: Those women at high risk for [bearing a child with] Down's syndrome can avoid that risk by cloning.

• Tay-Sachs disease: Sex-linked genetic disorders [such as Tay-Sachs disease] could be prevented by using cloning to ensure the sex of a baby and possibly could be cured.

• Liver failure: We may be able to clone livers for liver transplants.

• Kidney failure: We may be able to clone kidneys for kidney transplants.

• Leukemia: We should be able to clone the bone marrow for [transplants to treat] children and adults suffering from leukemia. This is expected to be one of the first benefits to come from cloning technology.

• Cancer: We may learn how to switch cells on and off through

cloning and thus be able to cure cancer. Scientists still do not know exactly how cells differentiate into specific kinds of tissue, nor do they understand why cancerous cells lose their differentiation. Cloning, at long last, may be the key to understanding differentiation and cancer.
• Cystic fibrosis: We may be able to produce effective genetic therapy against cystic fibrosis. Ian Wilmut and colleagues are already working on this problem.
• Spinal cord injury: We may learn to grow nerves or the spinal cord back again when they are injured. Quadriplegics might be able to get out of their wheelchairs and walk again. Christopher Reeve, the man who played Superman, might be able to walk again.
• Testing for genetic disease: Cloning technology can be used to test for and perhaps cure genetic diseases.

The above list only scratches the surface of what human cloning technology can do for mankind. The suffering that can be relieved is staggering. This new technology heralds a new era of unparalleled advancement in medicine if people will release their fears and let the benefits begin. Why should another child die from leukemia when, if the technology is allowed, we should be able to cure it in a few years' time?

Ethical Uses of Human Cloning

From e-mail to the Human Cloning Foundation (HCF), it is clear that many people would support human cloning in the following situations:

1) A couple has one child, but then they become infertile and cannot have more children. Cloning would enable such a couple to have a second child, perhaps a younger twin of the child they already have.

2) A child is lost soon after birth to a tragic accident. Many parents have written the HCF after losing a baby in a fire, car accident, or other unavoidable disaster. These grief-stricken parents often say that they would like to have their perfect baby back. Human cloning would allow such parents to have a twin of their lost baby, but it would be like other twins, a unique individual and not a carbon copy of the child that was lost under heartbreaking circumstances.

3) A woman through some medical emergency ends up having a hysterectomy before being married or having children. Such women have been stripped of their ability to have children. These women need a surrogate mother to have a child of their own DNA, which can be done either by human cloning or by in vitro fertilization.

4) A boy graduates from high school at age 18. He goes to a pool party to celebrate. He confuses the deep end and shallow end and dives head first into the pool, breaking his neck and becoming a quadriplegic. At age 19 he has his first urinary tract infection because of an indwelling urinary catheter and continues to suffer from them the rest of his life. At age 20 he comes down with herpes zoster [a viral infection] of the trigeminal nerve [a facial nerve]. He suffers chronic

unbearable pain. At age 21 he inherits a 10-million-dollar trust fund. He never marries or has children. At age 40, after hearing about Dolly being a clone, he changes his will and has his DNA stored for future human cloning. His future mother will be awarded one million dollars to have him and raise him. His DNA clone will inherit a trust fund. He leaves five million to spinal cord research. He dies feeling that although he was robbed of normal life, his twin/clone will lead a better life.

5) Two parents have a baby boy. Unfortunately, the baby has muscular dystrophy. They have another child and it's another boy with muscular dystrophy. They decide not to have any more children. Each boy has over 20 operations as doctors attempt to keep them healthy and mobile. Both boys die as teenagers. The childless parents donate their estate to curing muscular dystrophy and to having their boys cloned when medical science advances enough so that their DNA can live again, but free of muscular dystrophy.

Answers to Religious Arguments Against Human Cloning

Ronald A. Lindsay

Ronald A. Lindsay is a lawyer and philosopher whose specialty is bioethics. In the following essay, Lindsay counters several common religious objections to human cloning. For instance, he addresses the argument that cloning would rob people of their God-given uniqueness and dignity. According to Lindsay, there is no reason to believe that individuals produced through cloning technology will be less important to or valued by their loved ones than individuals born in the usual way. While Lindsay admits that abuses of human cloning are possible, he insists that religious precepts are insufficient to prevent such misuse. He maintains that the debate over human cloning should remain open, but that the arguments should be based on objective logic and reason rather than supernatural revelation.

The furor following the announcement of experiments in cloning, including the cloning of the sheep Dolly in February 1997, has prompted representatives of various religious groups to inform us of God's views on cloning. Thus, the Reverend Albert Moraczewski of the National Conference of Catholic Bishops has announced that cloning is "intrinsically morally wrong" as it is an attempt to "play God" and "exceed the limits of the delegated dominion given to the human race." Moreover, according to Reverend Moraczewski, cloning improperly robs people of their uniqueness. Dr. Abdulaziz Sachedina, an Islamic scholar at the University of Virginia, has declared that cloning would violate Islam's teachings about family heritage and eliminate the traditional role of fathers in creating children. Gilbert Meilander, a Protestant scholar at Valparaiso University in Indiana, has stated that cloning is wrong because the point of the clone's existence "would be grounded in our will and desires" and cloning severs "the tie that united procreation with the sexual relations of a man and woman." On the other hand, Moshe Tendler, a professor of medical ethics at Yeshiva University, has concluded that there is religious authority for cloning, pointing out that respect for "sanctity of life

Excerpted from "Taboos Without a Clue," by Ronald A. Lindsay, *Free Inquiry*, Summer 1997. Reprinted with permission.

would encourage us to use cloning if only for one individual . . . to prevent the loss of genetic line."

This is what we have come to expect from religious authorities: dogmatic pronouncements without any support external to a particular religious tradition, self-justifying appeals to a sect's teachings, and metaphor masquerading as reasoned argument. And, of course, the interpreters of God's will invariably fail to agree among themselves as to precisely what actions God would approve.

Given that these authorities have so little to offer by way of impartial, rational counsel, it would seem remarkable if anyone paid any attention to them. However, not only do these authorities have an audience, but their advice is sought out by the media and government representatives. Indeed, President Bill Clinton's National Bioethics Advisory Commission devoted an entire day to hearing testimony from various theologians.

Morality Does Not Need Religion

The theologians' honored position reflects our culture's continuing conviction that there is a necessary connection between religion and morality. Most Americans receive instruction in morality, if at all, in the context of religious belief. As a result, they cannot imagine morality apart from religion, and when confronted by doubts about the morality of new developments in the sciences—such as cloning—they invariably turn to their sacred writings or to their religious leaders for guidance. Dr. Ebbie Smith, a professor at Southwestern Baptist Theological Seminary, spoke for many Americans when he insisted that the Bible was relevant to the cloning debate because "the Bible contains God's revelation about what we ought to be and do, if we can understand it."

But the attempt to extrapolate a coherent, rationally justifiable morality from religious dogma is a deeply misguided project. To begin, as a matter of logic, we must first determine what is moral before we decided what "God" is telling us. As Plato pointed out, we cannot deduce ethics from "divine" revelation until we first determine which of the many competing revelations are authentic. To do that, we must establish which revelations make moral sense. Morality is logically prior to religion.

Moreover, most religious traditions were developed millennia ago, in far different social and cultural circumstances. While some religious precepts retain their validity because they reflect perennial problems of the human condition (for example, no human community can maintain itself unless basic rules against murder and stealing are followed), others lack contemporary relevance. The world of the biblical patriarchs is not our world. Rules prohibiting the consumption of certain foods or prescribing limited, subordinate roles for women might have some justification in societies lacking proper hygiene or requiring physical strength for survival. But they no longer have any utility and persist only as irrational taboos. In addition, given the limits of the world of the Bible and the Koran, their authors simply had no occasion to address some of the problems that confront us, such as the ethics of *in vitro* fertilization, genetic engineering, or cloning. To pretend otherwise, and to try to apply religious precepts by extension and analogy to these novel problems is an act of pernicious self-delusion.

Religious Objections to Cloning

To underscore these points, let us consider some of the more common objections to cloning that have been voiced by various religious leaders:

Cloning is playing god. This is the most common religious objection, and its appearance in the cloning debate was preceded by its appearance in the debate over birth control, the debate over organ transplants, the debate over assisted dying, etc. Any attempt by human beings to control and shape their lives in ways not countenanced by some religious tradition will encounter the objection that we are "playing God." To say that the objection is uninformative is to be charitable. The objection tells us nothing and obscures much. It cannot distinguish between interferences with biological process that are commonly regarded as permissible (for example, use of analgesics or antibiotics) and those that remain controversial. Why is cloning an impermissible usurpation of God's authority, but not the use of tetracycline?

Cloning is unnatural because it separates reproduction from human sexual activity. This is the flip side of the familiar religious objection to birth control. Birth control is immoral because it severs sex from reproduction. Cloning is immoral because it severs reproduction from sex. One would think that allowing reproduction to occur without all that nasty, sweaty carnal activity might appeal to some religious authorities, but apparently not. In any event, the "natural" argument is no less question-begging in the context of reproduction without sex than it is in the context of sex without reproduction. "Natural" most often functions as an approbative and indefinable adjective; it is a superficially impressive way of saying, "This is good, I approve." Without some argument as to why something is "natural" and "good" or "unnatural" or "bad," all we have is noise.

Cloning robs persons of their God-given uniqueness and dignity. Why? Persons are more than the product of their genes. Persons also reflect their experiences and relationships. Furthermore, this argument actually demeans human beings. It implies that we are like paintings or prints: the more copies that are produced, the less each is worth. To the contrary, each clone will presumably be valued as much by their friends, lovers, and spouses as individuals who are produced and born in the traditional manner and not genetically duplicated.

Beyond Theology

All the foregoing objections assume that cloning could successfully be applied to human beings. It is worth noting that this issue is not entirely free from doubt since Dolly was produced only after hundreds of attempts. And although in principle the same techniques should work in humans, biological experiments cannot always be repeated across different species.

Of course, if some of the religious have their way, the general public may never know whether cloning would work in humans, as research into applications of cloning to human beings could be outlawed or driven underground. This would be an unfortunate development. Quite apart from the obvious, arguably beneficial, uses of cloning, such as asexual reproduction for those incapable of having children through sex, there are potential spinoffs from cloning research that could prove extremely valuable. Doctors, for example, could develop techniques to take skin cells from someone with liver disease, reconfigure them to function as liver cells, clone them, and then transplant them back into the patient. Such a procedure would avoid the sometimes-fatal complications that accompany genetically non-identical transplants as well as problems caused by the chronic shortage of available organs for transplant.

This is not to discount the potential for harm and abuse that would result from the development of cloning technology, especially if we also master techniques for manipulating DNA. If we are able to modify a human being's genetic composition to achieve a predetermined end and can then create clones from the modified genetic structure, we could, theoretically, create a humanlike order of animals that would be more intelligent than other animals but less intelligent and more docile than (other?) human beings. Sort of ready-made slaves.

But religious precepts are neither necessary nor sufficient for avoiding such dangers. What we require is a secular morality based on our needs and interests and the needs and interests of other sentient beings. In considering the example just given, it is apparent that harmful consequences to normal human beings could result from the creation of these humanoid slaves, as many could be deprived of a means of earning their livelihood. It would also lead to an enormous and dangerous concentration of power in the hands of those who controlled these humanoids. And, although in the abstract we cannot decide what rights these humanoids would have, it is probable that, as sentient beings with at least rudimentary intelligence, they would have a right to be protected from ruthless exploitation and, therefore, we could not morally permit them to be treated as slaves. Even domesticated animals have a right to be protected from cruel and capricious treatment.

Obviously, I have not listed all the factors that would have to be considered in evaluating the moral implications of my thought experiment. I have not even tried to list all the factors that would have to be considered in assessing the many other ways—some of them now unimaginable—in which cloning technology might be applied. My point here is that we have a capacity to address these moral problems as they arise in a rational and deliberate manner if we rely on secular ethical principles. The call by many of the religious for an absolute ban on cloning experiments is a tacit admission that their theological principles are not sufficiently powerful and adaptable to guide us through this challenging future.

I want to make clear that I am not saying we should turn a deaf ear to those who offer us moral advice on cloning merely because they are religious. Many bioethicists who happen to have deep religious convictions have made significant, valuable contributions to this field of moral inquiry. They have done so, however, by offering secular and objective grounds for their arguments. Just as an ethicist's religious background does not entitle her to a special deference, so too her religious background does not warrant her exclusion from the debate, provided she appeals to reason and not supernatural revelation.

GLOSSARY

blastomere   A cell in the very early stages of conception; it is capable of developing into a whole organism if separated from other cells.

cytoplasm   The body substance of a cell, in which the **nucleus** and other organelles (specialized cellular parts) are embedded.

differentiated cell   A mature cell that has become a particular type, such as a nerve cell or muscle cell; many of its genes are turned off.

DNA   Deoxyribonucleic acid; the substance of which genes are made.

embryo   An unborn organism in the early stages of development (in humans, up to approximately the eighth week after conception).

enucleated egg   An egg cell from which the **nucleus** has been removed; another nucleus may be inserted into such an egg to start the process of cloning by **nuclear transfer**.

fetus   An unborn organism in the later stages of development (in humans, the fetal stage begins approximately three months after conception).

fibroblast   A cell that forms connective tissue; this type of cell is easily grown in laboratories and may be used as a donor for cloning.

genetic imprinting   The modification of genes in a transplanted **nucleus** by the **cytoplasm** of the cell into which the nucleus was transplanted.

genome   An organism's complete collection of genes.

genotype   The genetic makeup of an organism.

germ cell   A sex cell (egg cell or sperm cell), which carries genetic information from one generation to the next.

in vitro fertilization (IVF)   A scientific technique in which sperm and egg cells are combined in a laboratory to conceive a new organism; this procedure has been used to help infertile couples have children.

mitochondrial DNA   DNA found in the mitochondria, which are the parts of a cell that produce energy. The mitochondrial DNA in an **enucleated egg** may influence the genes in a **nucleus** transplanted into the egg for cloning and thereby modify the resulting organism.

nuclear somatic transfer or **somatic cell nuclear transfer**   A technique for producing a clone in which the **nucleus** from a **somatic cell** (such as a **fibroblast**) is put into a resting state by depriving it of nutrients and then is inserted into an **enucleated egg** from another organism (usually of the same species). The two are then fused together by a jolt of electricity.

nucleus   The central body of a cell, which contains all of the cell's genes except those in the mitochondria.

oocyte   An immature egg cell.

pluripotent cell   A cell that is still sufficiently undifferentiated to be able to grow into any type of **somatic cell** but not undifferentiated enough to grow into a whole organism.

recombinant DNA   DNA that has been spliced and recombined in the laboratory, usually so that it contains genes from more than one species.

somatic cell   A **differentiated** body cell; that is, any mature cell except a **germ cell**. Genetic information in somatic cells normally is not transmitted to future generations.

stem cell A type of undifferentiated cell, found in **embryos** and possibly in small amounts in adults, that can develop into many or all types of tissue.

sunset clause A clause in legislation that requires it to automatically expire after a certain period of time.

superovulation A condition, usually produced by fertility drugs, in which a female produces more eggs at one time than she normally would.

totipotent cell A cell capable of growing into a complete, fertile organism.

transgenic Containing genes from more than one species, usually as a result of genetic engineering.

xenotransplantation Transplantation of tissues or organs from one species to another.

ORGANIZATIONS TO CONTACT

The editors have compiled the following list of organizations concerned with the issues presented in this book. The descriptions are derived from materials provided by the organizations. All have publications or information available for interested readers. The list was compiled on the date of publication of the present volume; the information provided here may change. Be aware that many organizations take several weeks or longer to respond to inquiries, so allow as much time as possible.

Access Excellence
(800) 295-9881
e-mail: aeinfo@gene.com • website: http://www.accessexcellence.org

Sponsored by the biotechnology company Genentech and the National Health Museum, Access Excellence is a national educational program that provides scientific information via the Internet. Its website includes educational resources and activities, news about biotechnology, and links to other websites related to biotechnology and genetics. Publications available through Access Excellence's links include *Cloning: A Special Report* and *Your Genes, Your Choices*.

American Society for Reproductive Medicine
1209 Montgomery Highway, Birmingham, AL 35216-2809
(205) 978-5000 • fax: (205) 978-5005
e-mail: asrm@asrm.org • website: http://www.asrm.org

The society is devoted to advancing knowledge and expertise in reproductive medicine and biology. It offers services and education to health professionals, patients, and media professionals. Its publications include patient education booklets, press releases, and Capitol Hill briefs about upcoming legislation.

American Society of Law, Medicine, and Ethics
765 Commonwealth Ave., Suite 1634, Boston, MA 02215
(617) 262-4990 • fax: (617) 437-7596
e-mail: aslme@bu.edu • website: http://www.aslme.org

Members of this professional society include attorneys, physicians, health care administrators, and others interested in the relationship between law, medicine, and ethics. The society acts as an impartial forum for discussion of issues such as genetic engineering. It also maintains an information clearinghouse and a library. The society publishes the quarterly journals *American Journal of Law and Medicine* and *Journal of Law, Medicine, and Ethics*.

BC Biotechnology Alliance (BCBA)
1122 Mainland St., Suite 450, Vancouver, BC V6B 5L1, CANADA
(604) 689-5602 • fax: (604) 689-4198
website: http://www.biotech.bc.ca

The BCBA is an association for producers and users of biotechnology. The alliance works to increase public awareness and understanding of biotechnology, including the awareness of its potential contributions to society. The alliance's publications include the bimonthly newsletter *Biofax* and the annual magazine *Biotechnology in BC*.

Biotechnology Industry Organization (BIO)
1625 K St. NW, Suite 1100, Washington, DC 20006
(202) 857-0244 • fax: (202) 857-0237
website: http://www.bio.org

BIO is the chief trade and lobbying organization for the biotechnology industry, representing biotechnology companies, academic institutions, and state biotechnology centers. It is committed to the socially responsible use of biotechnology and works to educate legislators and the public about the nature and importance of biotechnology through its educational activities and workshops. Its publications include *Citizens' Guide to Biotechnology*, the bimonthly newsletter *BIO Bulletin*, and the biannual magazine *Your World, Our World*.

Council for Responsible Genetics
5 Upland Rd., Suite 3, Cambridge, MA 02140
(617) 868-0870 • fax: (617) 491-5344
e-mail: crg@essential.org • website: http://www.gene-watch.org

The council is a national organization of scientists, public health professionals, trade unionists, and others who work to ensure that biotechnology is developed safely and in the public interest. Its areas of concern include discrimination on the basis of genetic information, patenting of life-forms, food safety, and environmental quality. The council publishes *GeneWatch*, a bimonthly bulletin dedicated to the social implications of biotechnology, as well as educational materials and position papers on such subjects as the alteration of human genes.

Foundation on Economic Trends
1660 L St. NW, Suite 216, Washington, DC 20036
(202) 466-2823 • fax: (202) 429-9602
e-mail: FETcentury@aol.com • website: www.biotechcentury.org

The foundation examines emerging trends in science and technology and their impact on society, culture, the economy, and the environment. It believes society should use extreme caution when implementing genetic technologies to avoid endangering people, animals, and the environment. The foundation publishes the books *The Biotech Century* and *Reproductive Technology* as well as articles and research papers.

Genetics Society of America
9650 Rockville Pike, Bethesda, MD 20814-3889
(301) 571-1825 • fax: (301) 530-7079
e-mail: estrass@genetics.faseb.org •
website: http://www.faseb.org/genetics/gsa/gsamenu.htm

The society is a professional organization of scientists and academicians working in the field of genetic studies. It promotes the science of genetics and supports the education of students of all ages about the field. Its publications include the monthly journal *Genetics* and various educational materials on genetics and careers in genetic science, such as the brochures *The Science of Genetics: Solving the Puzzle* and *The Science of Genetics: Training and Careers*.

The Hastings Center
Garrison, NY 10524-5555
(914) 424-4040 • fax: (914) 424-4545
e-mail: mail@thehastingscenter.org • website: http://www.hastingscenter.org

This research institute addresses fundamental ethical issues in health, medicine, and the environment, including issues related to human genetics. It pursues interdisciplinary research and education that include both theory and practice and collaborates with policymakers in both the private and public spheres. It publishes the bimonthly *Hastings Center Report* as well as numerous books and papers on issues concerning biomedical ethics.

Human Cloning Foundation
PMB 143, 1100 Hammond Dr., Suite 410A, Atlanta, GA 30328
fax: (770) 396-3011
e-mail: HCloning@aol.com • website: http://www.humancloning.org

The foundation promotes education, awareness, and research about human cloning and other forms of biotechnology. It emphasizes the positive aspects of these new technologies. The foundation prefers to distribute its information over the Internet and requests that people refrain from contacting it directly for information; however, those who wish to contribute to the cause in some way are welcome to contact the foundation by mail. Its website offers a variety of resources, including essays on the benefits of human cloning.

Human Genetics Advisory Commission
Office of Science and Technology, Albany House, 94-98 Petty France, London SW1H 9ST, United Kingdom
(44) 0171 271 2131 • fax: (44) 0171 271 2028
e-mail: mileva.novkovic@osct.dti.gov.uk •
website: http://www.dti.gov.uk/hgac

The commission was established in 1996 to offer the British government independent advice on issues arising from developments in human genetics. Its publications on human genetics research, which are available on its website, include *Human Genetics: Learning for the Millennium and Beyond* and *Cloning Issues in Reproduction, Science, and Medicine.*

International Center for Technology Assessment
310 D St. NE, Washington, DC 20002
(202) 547-9359 • fax: (202) 547-9429
e-mail: office@icta.org • website: http://www.icta.org

The center analyzes the impacts of technology on society, many of which it considers to be negative. Its concerns include limiting genetic engineering (including human gene alteration), halting the patenting of life-forms and genes, and defending the integrity of food. The center endeavors to hold biotechnology companies responsible for their actions through legal petitions and litigation. One of its projects is Biotechnology Watch Human Applications, which monitors aspects of biotechnology that affect the human genome, including human cloning. The center's think tank, called the Jacques Ellul Society, publishes a newsletter and a journal.

Kennedy Institute of Ethics
PO Box 571212, Washington, DC 20057-1212
(202) 687-8099 • fax: (202) 687-8089
e-mail: kicourse@gunet.georgetown.edu •
website: http://www.georgetown.edu/research/kie

Established at Georgetown University in 1971, the Joseph and Rose Kennedy Institute of Ethics is a teaching and research center that offers ethical perspectives on major policy issues, especially in the biomedical field. Issues related to human genetics and gene alteration are among the subjects it considers. The institute's publications include the quarterly journal *Kennedy Institute of Ethics Journal,* an annual bibliography, and an encyclopedia. It also produces a series of papers that present overviews of issues and viewpoints related to particular topics in biomedical ethics.

National Bioethics Advisory Commission
6100 Executive Blvd., Suite 5B01, Rockville, MD 20892-7508
(301) 402-4242 • fax: (301) 480-6900
website: http://www.bioethics.gov

The commission was established in 1995 to advise the president of the United States about bioethics issues. It has considered such issues as human cloning and the rights of human subjects in research studies. Its publications include the extensive report entitled *Cloning Human Beings*.

National Human Genome Research Institute (NHGRI)
Bldg. 31, 31 Center Dr., MSC 2152, Bethesda, MD 20892
(301) 402-0911 • fax: (301) 402-0837
website: http://www.nhgri.nih.gov

Part of the government-sponsored National Institutes of Health, NHGRI oversees the Human Genome Project, the goal of which is to map the entire human genome. The institute's Ethical, Legal, and Social Implications (ELSI) Working Group addresses the public consequences of the project's research and the genetic information it generates. Both NHGRI and its ELSI Working Group sponsor research and education projects. NHGRI publishes reports about the Human Genome Project and a list of genomic and genetic resources on the Internet. The publications of the ELSI Working Group include press releases, fact sheets, and reports such as *Promoting Safe and Effective Genetic Testing in the United States*.

BIBLIOGRAPHY

Books

Lori B. Andrews — *The Clone Age: Adventures in the New World of Reproductive Technology*. New York: Henry Holt, 1999.

Ronald Cole-Turner, ed. — *Human Cloning: Religious Responses*. Louisville, KY: Westminster John Knox, 1998.

Rob DeSalle and David Lindley — *The Science of Jurassic Park and the Lost World: Or, How to Build a Dinosaur*. New York: Harper, 1997.

James C. Hefley and Lane P. Lester — *Human Cloning: Playing God or Scientific Progress?* Grand Rapids, MI: Fleming H. Ravell, 1998.

James M. Humber and Robert Almeder, eds. — *Human Cloning*. Totowa, NJ: Humana Press, 1998.

Leon R. Kass and James Q. Wilson — *The Ethics of Human Cloning*. Washington, DC: AEI Press, 1998.

Gina Kolata — *Clone: The Road to Dolly and the Path Ahead*. New York: Morrow, 1998.

George J. Marlin — *The Politician's Guide to Assisted Suicide, Cloning, and Other Current Controversies*. Dulles, VA: Morley Institute, 1998.

Gary E. McCuen, ed. — *Cloning: Science and Society*. Hudson, WI: Gary E. McCuen Publications, 1998.

Glenn McGee, ed. — *The Human Cloning Debate*. Berkeley, CA: Berkeley Hills Books, 1998.

Martha C. Nussbaum and Cass R. Sunstein, eds. — *Clones and Clones: Facts and Fantasies About Human Cloning*. New York: Norton, 1998.

Gregory E. Pence, ed. — *Flesh of My Flesh: The Ethics of Cloning Humans: A Reader*. Lanham, MD: Rowman & Littlefield, 1998.

M.L. Rantala and Arthur J. Milgram, eds. — *Cloning: For and Against*. Chicago: Open Court, 1999.

Melinda A. Roberts — *Child Versus Childmaker: Future Persons and Present Duties in Ethics and the Law*. Lanham, MD: Rowman & Littlefield, 1998.

Paul A. Winters, ed. — *Cloning: At Issue*. San Diego: Greenhaven Press, 1998.

Periodicals

Gary B. Anderson and George E. Seidel — "Cloning for Profit," *Science*, May 29, 1998.

George J. Annas — "Why We Should Ban Human Cloning," *New England Journal of Medicine,* July 9, 1998. Available from 10 Shattuck St., Boston, MA 02115-6094.

Ronald Bailey — "The Twin Paradox: What Exactly Is Wrong with Cloning People?" *Reason*, May 1997.

Sharon Begley — "Little Lamb, Who Made Thee?" *Newsweek*, March 10, 1997.

John Carey — "Barnyard Biotech Breeds High Hopes," *Business Week*, July 20, 1998.

Leon Eisenberg — "Would Cloned Humans Really Be Like Sheep?" *New England Journal of Medicine*, February 11, 1999.

Kathy A. Fackelman — "Cloning Human Embryos," *Science News*, February 5, 1994.

Mark Harris — "To Be or Not to Be?" *Vegetarian Times*, June 1998. Available from 4 High Ridge Park, Stamford, CT 06905.

Caryn James — "As Science Catches Up on Cloning, a Warning," *New York Times*, February 26, 1997.

Leon R. Kass — "The Wisdom of Repugnance: Why We Should Ban the Cloning of Humans," *New Republic*, June 2, 1997.

John F. Kilner — "Stop Cloning Around," *Christianity Today*, April 28, 1997.

Jeffrey Kluger — "Will We Follow the Sheep?" *Time*, March 10, 1997.

Gina Kolata — "For Some Infertility Experts, Human Cloning Is a Dream," *New York Times*, June 7, 1997.

Gina Kolata — "With Cloning of a Sheep, the Ethical Ground Shifts," *New York Times*, February 24, 1997.

E.V. Kontorovich — "Asexual Revolution," *National Review*, March 9, 1998.

R.C. Lewontin — "The Confusion over Cloning," *New York Review of Books*, October 23, 1997.

Eliot Marshall — "Biomedical Groups Derail Fast-Track Anticloning Bill," *Science*, February 20, 1998.

David Masci — "The Cloning Controversy," *CQ Researcher*, May 9, 1997. Available from 1414 22nd St. NW, Washington, DC 20037.

Jim Motavalli and Tracey C. Rembert — "Me and My Shadow," *E: The Environmental Magazine*, July/August 1997.

J. Madeleine Nash — "The Age of Cloning," *Time*, March 10, 1997.

Elizabeth Pennisi — "Cloned Mice Provide Company for Dolly," *Science*, July 24, 1998.

Elizabeth Pennisi and Nigel Williams — "Will Dolly Send in the Clones?" *Science*, March 7, 1997.

Jeremy Rifkin — "Dolly's Legacy: The Implications of Animal Cloning," *Animals' Agenda*, May/June 1997.

John A. Robertson — "Human Cloning and the Challenge of Regulation," *New England Journal of Medicine*, July 9, 1998.

Joan Stephenson — "Threatened Bans on Human Cloning Research Could Hamper Advances," *JAMA*, April 2, 1997. Available from 515 N. State St., Chicago, IL 60610.

Chana Freiman Stiefel   "Cloning: Good Science or Baaad Idea?" *Science World*, May 2, 1997. Available from 555 Broadway, New York, NY 10012-3999.

John Travis   "A Fantastical Experiment: The Science Behind the Controversial Cloning of Dolly," *Science News*, April 5, 1997.

Lindsay Van Gelder   "Hello, Dolly, Hello, Dolly," *Ms.*, May/June 1997.

Allen Verhey   "Theology After Dolly: Cloning and the Human Family," *Christian Century*, March 19, 1997.

Gurney Williams III   "Altered States," *American Legion*, October 1997. Available from PO Box 1055, Indianapolis, IN 46206-1055.

Christopher Wills   "A Sheep in Sheep's Clothing?" *Discover*, January 1998.

Ian Wilmut   "Dolly's False Legacy," *Time*, January 11, 1999.

Ian Wilmut et al.   "Viable Offspring Derived from Fetal and Adult Mammalian Cells," *Nature*, February 27, 1997. Available from Porters South, 4 Crinan St., London N1 9XW, United Kingdom.

Frank R. Zindler   "Spirits, Souls, and Clones: Biology's Latest Challenge to Theology," *American Atheist*, Summer 1997.

INDEX

ABS Global, 55
Address, Richard, 134
Advanced Cell Technology, 30, 43, 54, 55
Agnes, Michael, 132
agriculture, 63, 64, 65
  see also livestock industry
AIDS, 89
  vaccine for, 32
Alexion Pharmaceuticals Inc., 55, 57, 131
Alzheimer's disease, 153
American Intellectual Property Association, 107
American Livestock Breeds Conservancy, 64
American Medical Association, 107
Andrews, Lori, 37
Animal Scientific Procedures Act (Britain), 62
Animal Welfare Act, U.S., 62
animals, 62
  and medical research, 32
  selective breeding of, 29
  source of drugs, 14, 61, 82, 106
    current inefficiency of, 59
    interest of pharmaceutical companies in, 54, 55
    from protein in bodily fluids, 49
    and research problems, 56, 57–58
  source of organs for humans, 24, 49, 53, 60, 131
  see also cows; endangered species; livestock industry; pigs
Aquinas, Saint Thomas, 138
Aristotle, 138
Armey, Dick, 122
Asian elephant, 71, 72, 74
Asian horse, 67
Audubon Center for Research of Endangered Species, 69
Australia, 111

Bannon, Lisa, 84
Beltsville, Maryland, 65
Benirschke, Kurt, 66, 67, 68, 69
Bereano, Philip, 133
Bible, the, 93, 138–39, 140, 158, 159
birth control, 159
Bixby, Don, 64
Bohlin, Raymond G., 88
Bond, Christopher "Kit," 118, 123

Bondioli, Kenneth, 55, 56, 57
Boston Globe (newspaper), 45
Bouchard, Thomas, 37
Boys from Brazil, The (movie), 9, 19, 85, 129
Brave New World (Huxley), 9, 79, 81, 83
Briggs, Robert, 27
Bromhall, Derek, 79
Brookfield Zoo (Illinois), 67, 68
Bruford, Michael, 67

Callahan, Sidney, 136
Campbell, Keith, 21, 53, 55, 60
  key insight of, 19, 97
  on limitations of cloning technology, 20, 61
cancer, 154–55
Caplan, Arthur, 102, 110
Capron, Alexander Morgan, 107
Carr, Elizabeth, 45
Center for Bioethics, 102
Center for Reproduction of Endangered Species (CRES), 66, 67, 69
Chiba Livestock Experimentation Center, 50
Chicago Tribune (newspaper), 103, 104
Cimons, Marlene, 108
Clark, John, 96
Clinton, Bill, 9, 43, 82, 86, 158
  and ban on federal funding for human cloning research, 113
  on possible benefits of cloning research, 111
  and threat to humanity, 122
  and request for review of implications of Dolly's creation, 106, 112, 120, 129, 144
  speech by, 108–109
Cloning of Joanna May, The (Weldon), 81
Coe, Christopher, 37, 38
Cohen, Jon, 66
Colman, Alan, 54, 57, 97, 98
Combes, Bob, 60
Congress, U.S., 31, 65, 96, 106, 107
  ban on federally funded human embryo research, 109, 125
  forbidden from legislating on religion, 123
Conservation and Research Center, 67
Constitution, U.S., 107, 123

Consumer Reports (magazine), 120
Cornett, Walter G., III, 102
Council for Science and Technology, 49, 51
Coupland, Douglas, 83
cows, 43, 64, 73
  cloning of, 25, 29, 30, 35
    from cells of single fetus, 33
    for production of prized breeds, 32
creation, 93
Crichton, Michael, 88, 89, 91, 93
Critser, John, 72
Cryobiology Research Institute, 72
cystic fibrosis, 40, 59, 61, 122, 155

Dallas Morning News (newspaper), 93
Davidson, Keay, 63
Deegan, Robert Cook, 125
Department of Agriculture, U.S. (USDA), 63, 64, 65
  Animal and Plant Health Inspection Service, 61
diabetes, 40, 41, 153
DiBerardino, Marie A., 27, 29
Dickey, Jay, 43
Die Hard (movie), 85
Dingell, John D., 110
DNA, 16, 24, 27, 58, 61
  can be stored for future cloning, 156
  human, mingled with animal cells, 43
  key ingredient in cloning, 34
  modification of reversible, 29
  and reprogramming of cells, 31
Dolly (cloned sheep), 19, 30, 100, 109, 129
  adult cell DNA fused with donor egg to create, 43, 48, 54, 74
    after multiple attempts, 10, 21, 36, 68–69, 160
  inspired by casual rumor, 16
  to produce proteins for drugs, 14
  repetition of, not planned, 20, 98
  though previously considered impossible, 31
  called Frankenstein's monster, 26
  importance of, 13, 27, 31, 53
  media reactions to, 82, 85, 86, 98, 99
  named after Dolly Parton, 9, 20
  questions generated by, 15, 22, 36, 129–30
    because of hoaxes in history, 23–24
    and cautious reactions, 108–109, 113

including issue of human cloning, 112, 120
Down's syndrome, 39, 122, 154
Dresser, Betsy, 69, 70
Duvannyi Yar, 75, 76

Earth in the Balance (Gore), 65
Easterbrook, Gregg, 39
Ehlers, Vernon, 119
Eibert, Mark D., 118
Einstein, Albert, 13
Elshtain, Jean Bethke, 9, 145
embryos, 40, 90, 109
  cross-species, 56
  division of, 33–35
  DNA can be used to support development of, 58
  and fetal tissue research, 90, 110
  and research, 138
  research on, banned, 41
  twinning in IVF, 126
  surplus of, from IVF clinics, 41, 42
endangered species, 49–50, 73
  difficulty of cloning, 69
  possible disadvantages of cloning, 68
  potential of cloning to save, 67, 70
  program to freeze cells from, 66
Engelhardt, Dean L., 28
Enzo Biochem Inc., 28
eugenics, 121–22, 137, 151
European Community, 107
evolution, 91
  cloning a threat to, 141–43
extinct animals
  re-creation of, impossible, 93, 94
  search for remains of, 71, 73
    in Siberian permafrost, 72, 75
      for cloning, 74, 77
      frustrations of, 76
Eyestone, Will, 54

families, 143, 148–49
  human cloning a threat to, 114, 132, 138, 139
  and parenthood, 152
Feinstein, Dianne, 118, 119
Feldmann, Eddie, 85
fertility drugs, 122, 126, 127
fictional accounts of cloning, 79, 129
  mixed with fact, 80
  used in publicity, 81–83
  see also movies
Field, Loren, 42
Firestone, Shulamith, 137
First, Neal, 18, 25, 56, 57, 58
Fleischman, Alan R., 27

Fletcher, David, 134
Follett, Ken, 86
Food and Drug Administration (FDA), 61, 119
Food Technology Magazine, 64
Foote, Robert, 62
Foundation on Economic Trends, 106
Frankenstein (Shelley), 9, 14, 83
Fund for the Replacement of Animals in Medical Experiments (FRAME), 60

Gattaca (movie), 86
Gearhart, John, 40, 41
    call for legal guidelines by, 42
    Down's syndrome researcher, 39
Gellman, Rabbi Marc, 28
genetic diversity, 67, 74, 136
    cloning a threat to, 141–42
    need for, 65
    reduction in, 63
    risks of, 64
genetic engineering, 149–50, 152, 159
    see also animals; human cloning
Genzyme Transgenics, 55, 133
Germany, 126
Geron, 41, 42, 43
giant panda, 67
Goldfine, Phillip, 85
Goodman, Roy M., 27
Gordon, Carl, 61
Gordon, Meg, 59
Gore, Al, 65
Goto, Kazufumi, 71, 72, 74, 76
    and discovery regarding dead sperm, 73
    and promise to clone mammoth, 77
Granada Genetics Inc., 56, 62
Griffin, Harry, 86–87, 98, 100
Gurdon, John, 18

Hall, Jerry, 81
Harkin, Tom, 43
Herbert, Wray, 129
Hogan, Brigid, 18
Holmes, Oliver Wendell, 121
Hoppe, Peter, 18
Hotz, Robert Lee, 95
human cloning, 37, 53, 82, 99
    ban on, 118, 144
        in California, 119
        in conflict with constitutional rights, 120
        in European countries, 129
        inappropriateness of, 125–27, 147
    debate on, 106, 148
        and comparison to eugenics, 121–22

    need for, 111, 117, 144
    ethical benefits of, 155–56
    ethical concerns about, 45, 114, 134, 138–40
        can be addressed through secular morality, 160–61
        and issues of power, 137
    family relationships, 148–49
    fears about, are unrealistic, 99, 124, 145
    federal funding ban, 96, 107, 108–109, 110, 111
    likely to be rare, 152
    possibility of, 30, 36–37, 112
        as hope for infertile couples, 38
        is remote, 31
        negative public reaction to, 135
        caused by speed of development, 46
    possible legislation about, 115–16
    religious objections to, 45
        irrelevance of, in modern world, 158–59
        similar to early attitudes about transplants, 123
    risks of, include, 132
        genetic defects, 138
        loss of personal uniqueness, 136
        social problems, 133, 146
        and loss of role for fathers, 137, 157
    and stem cell research, 44, 46
    uniqueness of individuals not threatened by, 37–38, 130–31, 145
    see also families; National Bioethics Advisory Commission (NBAC)
Human Cloning Foundation, 153
Human Embryo Panel, 125
Human Genome Project, 127
Huxley, Aldous, 9, 79, 80, 81, 83

Illmensee, Karl, 18, 19, 23
In His Image (Rorvik), 79
infertility, 38, 120, 154, 155
    and dangerous fertility drugs, 122, 126–27
Institute of Food Technologies, 64
Institute of Zoology (London), 67
Institutional Review Boards (IRBs), 126
International Center for Technical Assessment, 60
International Embryo Transfer Society, 53, 55, 58
Iritani, Akira, 74, 75
IVF (in vitro fertilization), 29, 41, 45, 111, 155

babies born through, 123
embryo twinning in, 126–27
and experiments to produce
multiple embryos, 81
legalized in spite of risks, 122
as treatment for infertile couples,
147–48

Jacobs, Margaret A., 106
Japan, 49, 52, 72, 73, 74
cloning experiments in, 51
genetic engineering for livestock of,
50
Johnson, Dirk, 101
Jonsen, Albert, 133
Jurassic Park (movie), 88, 93, 94
on dangers of science, 89–90
marginalization of religion by, 91,
92

Kagoshima Prefecture Cattle Breeding
Improvement Research Institute, 51
Kagoshima University, 72
Karloff, Boris, 26
Kass, John, 103
Kass, Leon R., 44, 45
Katz, Frances, 64
Keaton, Michael, 10, 85–86
Kelly, Chris, 85
Kennedy, Ted, 118, 119
Kennedy Institute, Georgetown
University, 62
Kimbrell, Andrew, 60
King, Thomas J., 27
Kinki University, 48, 74
Kinoshita, Akihiro, 73
Kleinberg, Lewis, 84
Kobayashi, Kazutoshi, 73, 74, 75, 76
Kolata, Gina, 13

Lacy, Robert, 67, 68
Last Harvest, The (Raeburn), 65
Layman, Lawrence, 36
Lazarev, Pyotr, 74
Leakey, Julian, 36
Ledbetter, David, 30, 36
leukemia, 40, 154
Lewis, C.S., 137
Lewontin, Richard, 146
Lindsay, Ronald A., 157
livestock industry, 48, 49, 100
and development of cloning
techniques, 51–52
in Japan, 50, 72
see also animals; Dolly (cloned
sheep)
Lutz, Diana, 26

Maher, Bill, 85
Mahon, Geoff, 16
Marchi, John J., 27
marriage, 138–39
Massachusetts University, 49
Mautner, Michael, 141
McCarthy, Charles, 62
McCormick, Richard, 137
McDermott, Anne, 85
McGrath, James, 18, 19
McKay, Ron, 41
McKernan, Ruth, 83
media, 85, 93, 100, 101
portrayal of Ian Wilmut by, 96
reaction to organ transplantation,
99
and Richard Seed, 102, 103, 104
Meilaender, Gilbert, Jr., 150, 157
mice, 18, 42, 43, 73
Ministry of Agriculture, Forestry, and
Fisheries (MAFF), 50, 51
Mitalipova, Maissam, 58
monkeys, 30, 38, 98, 108, 109
cloning of, 31
by embryo division, 34–35
likely to remain rare, 33
new lab for, 32
Moraczewski, Albert, 123, 157
Morell, Virginia, 30
movies, 84–87
Multiplicity (movie), 10, 85–86, 129
muscular dystrophy, 122, 156

National Academy of Science, 125
National Advisory Board on Ethics in
Reproduction, 131
National Bioethics Advisory
Commission (NBAC), 107, 126, 149,
151
headed by Harold T. Shapiro, 11
report on human cloning, 117
and consultation with wide
variety of experts, 113
including theologians, 123, 158
criticized for deference to
opponents, 146
and examination of religious
traditions, 114, 122
public policies recommended by,
115, 125
requested by President Clinton,
10, 106, 109, 120, 144
National Cancer Institute, 22
National Center for Toxicological
Research, 36
National Conference of Catholic
Bishops, 123

National Federation of Agricultural
  Cooperative Associations, 50
National Institute of Animal Industry,
  50, 51
National Institute of Neurological
  Disorders and Stroke, 41
National Institutes of Health (NIH), 30,
  40, 42, 107, 127
  provider of research grants, 32, 33
    human cloning experiments not
      supported by, 109
  Recombinant DNA Advisory
    Committee (RAC), 126
National Public Radio (NPR), 103
National Zoo, U.S. (Washington, D.C.),
  67
Nature (magazine), 17, 22, 69, 81, 82
Netherlands, the, 55
New Republic (magazine), 9
Newsweek (magazine), 81, 82, 83, 93
New York Academy of Medicine, 27
New York Academy of Sciences, 26
New York Times (newspaper), 13, 81, 89
nuclear transfer techniques, 50, 55,
  125–27, 146
  first experiments in, 18
  and nuclear embryo transfer, 31,
    33–35, 57
  and ongoing research, 113
  rapid advances in, 56
  to replicate positive traits in
    livestock, 48–49
Nurse, Paul, 83

Oregon, 108, 109
Oregon Regional Primate Research
  Center, 31
organ transplantation, 53, 103, 123,
  131, 154
  and assumption that human clones
    will provide organs, 137–38
  see also animals

Palca, Joe, 103
Parens, Erik, 27
Parkinson's disease, 40, 153
Patent and Trademark Office, U.S., 107
Pence, Gregory E., 125
Pennisi, Elizabeth, 11, 53
Peterson, Jonathan, 108
Pharming Holding N.V., 55
Philadelphia Daily News (newspaper), 9
Piedrahita, Jorge, 24
pigs, 61, 62, 64, 131
  and biomedicine, 55, 56
Piltdown man, 23
Pleistocene era, 76

population growth, 142
Post, Stephen G., 135
PPL Therapeutics. See Roslin Institute
Prather, Randall S., 18, 22, 56, 130

Radio Times (magazine), 80
Raeburn, Paul, 65
Rea, Teresa Stanek, 107
Reagan, Ronald, 110
reconstructive/cosmetic surgery, 154
Rissler, Jane, 64
Robertson, John, 131–32
Robl, James M., 19, 24, 55, 56, 58
  on pharmaceutical industry's
    interest in cloning, 54
Rorvik, David, 79, 80
Rose, Frederick, 84
Roslin Institute, 19, 49, 74, 85, 112
  animal research laboratory in
    Scotland, 14, 22–23
  and cloning research
    high success rate in, 121
      including Dolly the sheep, 10–11
    as model for other scientists, 58
  and PPL Therapeutics, 53, 59, 60, 97,
    133
  cows, as well as sheep produced
    by, 55
  and failures in transgenic cloning
    experiments, 56, 57
  patent for cloning technology
    sought by, 24
  and time spent on egg injection
    research, 54
  and worldwide press attention,
    86–87
Rosner, Fred, 138
Roth, Joe, 86
Russia, 73
  see also Siberia
Ryder, Oliver, 66, 67, 68, 69
Ryuzo, Yanagimachi, 48

Sachedina, Abdulaziz, 157
San Diego Zoo, 66
Sauer, Mark, 11
Scarr, Sandra, 37
Schecter, Roger E., 107
Schramm, Dee, 32, 33, 34
Schroeder, Beth, 86
Science (magazine), 11, 18
Science and Technology Agency, 49
Seed, Randolph, 104
Seed, Richard, 31, 53, 101–104
Seed, The, (movie), 84
Seidel, George, 56, 61
Seiya, Takahashi, 48

Sgaramella, Vittorio, 57
Shapiro, Harold T., 11, 112
Sheler, Jeffery L., 129
Shelley, Mary Wollstonecraft, 9, 14, 26
Siberia, 74, 77
    search for woolly mammoth
        remains in, 71
    frustrations of, 72, 75, 76
sickle-cell anemia, 40
Silver, Lee, 37, 38
Smith, Ebbie, 158
Solter, Davor, 18, 19
Soule, Michael, 68
South China tigers, 67
Spanish ibex, 67
Specter, Arlen, 43
Specter, Michael, 13
Spemann, Hans, 17, 18, 35
Spielberg, Steven, 88, 89, 93
Stackhouse, Max, 134, 139
Stanford University Medical Center, 123
Star Wars (movie), 92
stem cell research, 39
    and difficulty in obtaining materials,
        41
    and link to human cloning, 45
    medical benefits of, 40, 44, 153–55
    need for public funding of, 43
    need to regulate, 42, 46
Stewart, Colin, 22
Stice, Steven, 35, 55, 58
Stone, Richard, 71
Supreme Court, U.S., 107, 120, 121
surrogacy, 151, 155

Tadashi, Sugie, 50
Tang, Frank, 61
Taubes, Gary, 23
Tay-Sachs disease, 122, 154
Tendler, Moshe, 157–58
Thomson, James, 40
    on need for regulation of embryo
        research, 42
    use of IVF embryos by, 41
Time (magazine), 81, 82–83
Turner, George, 80
Turney, Jon, 79
twins, 37, 38, 44, 131, 132

U.S. News & World Report (magazine),
    129, 130
Union of American Hebrew
    Congregations, 134
Union of Concerned Scientists, 64
University of Wisconsin, Madison, 56,
    57, 58

University of Massachusetts, Amherst,
    58
University of Missouri, Columbia, 56
USA Today (newspaper), 93

Varmus, Harold, 40, 110, 111
Verhey, Allen, 9–10

Wachbroit, Robert, 144
Waldholz, Michael, 22
Wall, Robert, 24
Wall Street Journal (newspaper), 89
Washington Post (newspaper), 149
Watkins, David, 33
Watson, Traci, 129
Webster's New World Dictionary, 132
Westhusian, Mark, 57
Wildes, Kevin, 134
Wildlands Project, 68
Wildt, David, 67, 68
Willadsen, Steen M., 16, 17, 36, 98
Wilmut, Ian, 28, 29, 53, 109, 133
    career of, 15–16, 17, 98
        as embryo scientist, 22, 23
        and innovation that led to birth
            of Dolly, 31, 36, 43, 74, 129
        and working environment, 97
    on need for regulation against
        human cloning, 69, 99
    news of success withheld by, 24
    as private man, 96
        disturbed by publicity, 95, 100
        and reluctance to acknowledge
            implications of his work,
            13–14, 21, 60
Wilson, James Q., 145
Wisconsin, 20, 121
Wistar Institute, 18
Wolf, Don, 31, 34, 36
    and monkey cloning project, 32, 33,
        35
woolly mammoth, 71, 72, 74, 75, 77

xenotransplantation, 49

Yakobian, Stephen, 84
Yakutia, 74
Youngs, Curtis, 23
Yukio, Tsunoda, 48, 50, 51

Zgorski, Lisa-Joy, 107
Zimov, Sergei, 71, 75, 76, 77
Zinder, Norton, 57
Ziomek, Carol, 55
Zoo Biology (journal), 68